Napoleon's Last Victory

NAPOLEON'S LAST VICTORY AND THE EMERGENCE OF MODERN WAR

Robert M. Epstein

Foreword by Russell F. Weigley

University Press of Kansas

Published by the University Press of Kansas (Lawrence, Kansas 66049), which was organized by the Kansas Board of Regents and is operated and funded by Emporia State University, Fort Hays State University, Kansas State University, Pittsburg State University, the University of Kansas, and Wichita State University

Library of Congress Cataloging-in-Publication Data

Epstein, Robert M.
    Napoleon's last victory and the emergence of modern war / Robert M. Epstein ; foreword by Russell F. Weigley.
       p.   cm.—(Modern war studies)
    Includes bibliographical references and index.
    ISBN 0-7006-0664-5
    1. Wagram, Battle of, 1809. 2. Napoleon I, Emperor of the French, 1769–1821—Military leadership. 3. Napoleonic Wars, 1800–1815—Campaigns—Austria. 4. Military art and science—France—History—19th century. 5. Strategy—History—19th century.
    I. Title. II. Series.
    DC234.8.E67   1994
    940.2′7—dc20   93-38243

Maps 2–6, 9–12, and 14–16 (Campaign of 1809 maps) are reproduced courtesy of the Department of History, U.S. Military Academy, West Point, N.Y.

British Library Cataloguing in Publication Data is available.

Printed in the United States of America
10 9 8 7 6 5 4 3 2 1

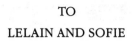

TO
LELAIN AND SOFIE

CONTENTS

MAPS

KEY TO SYMBOLS

SIZE OF UNIT

Army		Brigade	x
Corps	xxxx	Regiment	ꞏ ꞏ ꞏ
Division	xxx	Battalion	ꞏ ꞏ
	xx		

COMBAT ARM

Infantry   ⊠   Cavalry   ⊘

FOREWORD

Robert M. Epstein offers in these pages a stimulating return of military history to its roots. He brings us military history as primarily the study of the conduct of war.

During the nearly two generations since World War II, military history as written and studied in the United States has greatly and usefully expanded its vision to place war and armed forces in their political, economic, social, and cultural contexts; but in the process there has emerged a tendency sometimes to lose sight of the unpleasant but essential truths that armed forces exist principally to engage in combat (and also of course to deter combat, but waging war is their ultimate purpose) and that military history must therefore return continually to a focus on war. Given also the distasteful truth that warfare shows little sign of disappearing from our world, but may instead plague us yet more than it did while the Cold War balance between the two superpowers acted to deter at least some conflict, there is utility also in regarding the history of past warfare as a source of instruction for the conduct of war in the future, as Epstein must do as a professor in the U.S. Army School of Advanced Military Studies and as he does in transferring his classroom approach to the pages of this book.

He instructs us particularly in these pages on the emergence during the later Napoleonic Wars, and specifically during the campaign of 1809, of one of the salient features distinguishing modern warfare from wars whose characteristics were those of an age now past. We might debate endlessly, and for the most part fruitlessly, just when it was that war became altogether modern, but Epstein assuredly demonstrates that an especially important aspect of military modernity did appear in the late Napoleonic campaigns. That aspect is distributed maneuver, the deployment across a theater of military operations of military formations and units so constructed that each of them was capable of operating to a large extent independent of the others. More than any other factor the articulation of military forces into numerous autonomous entities is responsible for the extraordinary resiliency of modern military forces, their ability to bear up under immense punishments inflicted by their enemies, their often amazing capacity to avoid destruction. In well-articulated armies, a division or two may fall, but the autonomy of the various divisions and corps makes it almost impossible for the whole army to be destroyed.

Epstein explores the emergence of distributed maneuver and its corollary of articulated formations in the context of the considerable expansion of the size of armies in the later Napoleonic Wars. In order to administer and maneuver effectively forces that increasingly numbered several hundreds of thousands of soldiers, there had to occur by 1809 not only a continual improvement in the *corps d'armée* system of organization above the level of divisions but also a further evolution toward multiple field armies, each comprising a number of corps, and even toward groups of armies. Although Epstein's emphasis is on autonomous articulation of large formations, most of what he says about the nature, value, and importance of distributed maneuver in making armed forces more resilient in the face of their enemies' efforts to destroy them has implications also for a possible warfare of the future in which we may deal with armed forces on a lesser scale, perhaps in arenas opened up by the disintegration of the Soviet Union and its empire, but in which the intractability of autonomous units in the face of their would-be destroyers will remain a critical element.

Especially pertinent to understanding military operations as conducted through the distributed maneuver of an army's autonomous subdivisions is Epstein's emphasis on the rise of military commanders subordinate to the highest authority but nevertheless educated to operate independent of that authority when circumstances might make doing so appropriate. Formations trained to maneuver independently would have been of little use without commanders prepared to think and act on their own responsibility. Emperor Napoleon may not have pursued as far as he should have the logic whereby the corps and field army systems pointed to the nurturing among the marshalate and his other senior generals of an educated, professional capacity and willingness to exercise independent command. Nevertheless, Napoleon certainly displayed at least some appreciation of the growing necessity to have commanders willing to judge and act for themselves. In addressing this issue, Epstein explores the particular pertinence of the teacher-pupil relationship between Napoleon and his Viceroy of Italy and Colonel General of the Chasseurs of the Guard, Prince Eugene de Beauharnais. Eugene's absorption of Napoleon's instruction in the profession of arms and in the principles of independent military command lies at the heart of this chronicle of the campaign of 1809.

This book is a chronicle of warfare in a colorful, stirring era. Those of us addicted to reading about the pageantry and spectacle of war, however much we may reprimand ourselves for thereby neglecting war's horrors, can savor Robert Epstein's book as another entryway to the drama of war in the age of Napoleon. But we can find in this book, too, a specimen of combat history in an instructive mode to prepare military professionals for possible tasks yet to come.

Russell F. Weigley
Philadelphia, Pennsylvania

ACKNOWLEDGMENTS

I wish to thank the many people without whom the writing and publication of this book would have been more difficult, if not impossible. My thanks go to Gunther E. Rothenberg, Dennis E. Showalter, Russell F. Weigley, and Scott Bowden, all of whom were willing to take the time to read and provide wise counsel on the manuscript. Their contributions challenged me and kept me on the right track. In addition, I wish to acknowledge the intellectual support and guidance of Jacob W. Kipp, James L. Schneider, and Michael Briggs. I also thank Colonel Robert A. Doughty of the Department of History at the U.S. Military Academy at West Point for granting permission to use the maps titled "Campaign of 1809" in this book. I would also like to thank Edward Krasnoborski of the Department of History of the U.S. Military Academy at West Point, who did the artwork for those maps. In addition, I wish to thank Harry L. Fink for drawing the maps for the operations in Italy and Hungary. I owe a debt of gratitude to David Chandler whose monumental *The Campaigns of Napoleon* inspired me to become interested in Napoleonic history. Finally, I wish to thank my wife, Lelain, for the countless hours she spent as my personal editor on this manuscript. I offer my heartfelt thanks to all of the people mentioned above for their contributions. Any mistakes found in this book are mine alone.

Introduction

What is the purpose of studying the history of warfare? Some people would argue that there is no purpose at all; others might say that by studying wars, one helps to perpetuate them. I will not debate the morality of studying this particular subject. Warfare is studied academically, and military history is certainly popular as any perusal of a bookstore or movie listing will show. For many people who teach or study this subject, the reasons vary, but often there is a strong element of personal interest.

For me, the issue goes beyond personal interest or intellectual curiosity, because I am a professor of history at the U.S. Army's School of Advanced Military Studies in Leavenworth, Kansas. The study of history constitutes one-third of the curriculum and plays a major role in the development of key staff planners in the U.S. Army. I take teaching very seriously, and my major concern is to avoid the misuse of history. I sometimes awaken in the middle of the night fearful that my name may someday be linked to that of Grandmaison or Foch, French officers who believed they had uncovered the secrets of military success through the study of history. They thought history proved that principles of war were constant and manifest during the Napoleonic era and developed theories and doctrine based on that belief. Thousands of men were slaughtered in World War I, in part because of this misuse of history and its pairing with military doctrine.[1] How do I, as a teacher of officers, avoid making a similar mistake? Should one not teach history at all to impressionable student officers and so avoid the danger entirely? That is impossible. Officers and military institutions spend most of their time at peace, and the past is one of the few places where one can study warfare to prepare for the terrible day when it occurs again. Officers and interested civilians study war on their own, very often the lives of great commanders, in a search for role models to unlock the secrets of leadership. Yet their study can be uncritical or, worse, narrowly selective. Popular interpretations of history exalt heroic actions and so reinforce the myth that genius can conquer all. The context of individual actions is often lost in these drum-and-trumpet accounts.

History provides a reference point for anyone seeking to learn about war. However, establishing a reference point or a means of comparison is difficult because of the advent of rapid technological change. In earlier days it was easier to believe that there were constant factors in war. A soldier at Blenheim would fit in easily at Waterloo, but a trooper from Waterloo would find a battle in 1915 incomprehensible. Exact comparisons between wars of different technological periods cannot be made, but that does not mean that the usefulness of history fades.

I see two purposes for studying history, one negative and the other positive. The negative purpose is that history needs to be studied to challenge or disprove military theories or doctrines. It is most difficult to understand that the only real lesson from history is that there are no overall theories or doctrines, not any scientific ones at least. Each case is unique; there are too many variables to try to reduce history to a set of rules or lessons learned. In war, events are determined in a dynamic struggle between two forces. Success is completely relational. One only has to be better than the opponent to win. History is evolutionary, not static, and the same can be said about warfare. An understanding of this dynamic is one way to avoid misusing history, and this understanding is to be gained, as Michael Howard says, by studying the subject in breadth, depth, and context.[2]

The positive purpose for studying history is that by doing so in depth, one can learn what is important in the conduct of war and what the actual elements of war, or its characteristics, are in a given time or place. An understanding of the dynamics of a given situation might lead to the mastering of that situation. Learning to look at the fundamentals will give meaning to experience and so help a student understand the present conduct of war.

The issues of change and evolution are major themes in the history courses I teach. To show that situations are dynamic, not static, and to explain how individuals and institutions adapt or fail to adapt to change, I attempt to teach history in breadth. My teaching has led me to a comparative examination of military operations. I discovered that the orthodox interpretations of the evolution of warfare did not hold up, especially in the nineteenth century. Most histories treat the French Revolution and the Napoleonic Wars as a single block. Military theorists and historians such as Antoine Henri Jomini, Carl von Clausewitz, and B. H. Liddell Hart emphasize Napoleon Bonaparte's Italian campaigns of 1796–1797 and the glittering victories of 1805–1807 as models to be studied.[3] It was in these campaigns that great lessons could be drawn and theories produced. The figure of Napoleon dominates the period to such an extent that the fortunes of French arms 1796–1815 are often interpreted as the rise and fall of Napoleon's genius. Although changes in the nature of the opposing armies are mentioned, this development is often placed in the background of the drama.[4] In the course of my study, I found that there were significant differences in the way warfare was conducted during the years 1792–1815. The approach to military operations, the conduct of campaigns and battles, and the organization of armies changed a great deal during this period. Rather than a unitary block, the period seemed to divide into distinct phases: the laying of the foundations of Napoleonic warfare, 1792–1805; its dramatic emergence and years of dominance, 1805–1807; the Peninsular War that began in 1807; and the restoration of military equilibrium, 1809–1815.

Orthodoxy did not seem to stand up to scrutiny when my studies took me further into the nineteenth century. The usual interpretation was that technology significantly altered warfare. Railroads and the telegraph systemized strategic movements, and rifled weapons made Napoleonic frontal assaults obsolete, establishing the dominance of the tactical defense. The marvels of the machine age ensured that larger armies would be fielded and battlefronts expanded. However, this interpretation did not bear up under closer analysis. Technology most certainly changed warfare in the middle of the nineteenth century, but that change was not uniform throughout. The railroads and telegraph had a role in strategic deployment, but once the railheads were left behind, armies still marched on foot and foraged as in the early years of the century.

Armies of the mid-nineteenth century were organized into corps, which gave them endurance and resilience. They were difficult to destroy in a single battle unless they could be surrounded. If not surrounded, they would soon revive and the campaign would continue. The same seemed to hold true for Napoleonic armies after 1807. The size of the armies and the battlefronts of the wars of the mid-nineteenth century were similar to the size and frontages of the Napoleonic-era armies after 1809.

It was argued that rifled weapons in the mid-nineteenth century rendered frontal assaults suicidal. Yet the pattern of the tactical superiority of the defense seemed to be established in the latter half of the Napoleonic Wars, when assaulting infantry were blown away by combined fire of musketry and cannon. By 1809 successful bayonet attacks had become almost impossible against formed troops relying on firepower.

The picture that emerged when comparing the conduct of war in the latter half of the Napoleonic era with the wars of 1861–1871 seemed to be a blending of characteristics rather than a clear break with the past; I could ignore neither the differences nor the similarities. Warfare conducted in 1809–1815 had more affinity to the wars of 1861–1871 than to the first half of the Napoleonic Wars. If this blending of the period 1809–1871 is correct, then what are the implications for the origins of modern warfare?

What is modern warfare and where does it begin? Its starting point is difficult to determine. It can be argued that it began at the end of the Middle Ages with the advent of fire weapons. There are some historians who argue that modern war emerged only after 1871 when industrialization, technology, and the machine age fully developed. Others find the origins in the middle of the nineteenth century, very often in the American Civil War. Russell F. Weigley argues that strategic mobilization orchestrated and directed by Ulysses S. Grant during the American Civil War ushered in modern warfare.[5] Michael Howard makes a similar claim stressing strategic logistics in his "Forgotten Dimensions of Strategy."[6] Edward Hagerman in *The Civil War and the Origins of Modern Warfare* emphasizes the emergence of technology, especially the use of the rifle combined with entrenchments that created tactical deadlock on the battlefield, as the starting point.[7] Matthew Cooper in *The German Army* considers that modern war can be traced back to Moltke and the Wars of German Unification that witnessed the combination of mobilization and strategic deployment by railroad combined with an operational orientation toward envelopment.[8]

If the use of technology and industrialization is the critical issue in determining the emergence of modern war, then it is correct to place the origin in the middle of the nineteenth century. But by doing so one accepts the changes at the tactical and strategic levels as definitive, and this ignores the operational level.

Is technology the sum determinant of modernity or are there other factors to consider? What of the structural, organizational, intellectual, relational, or operational elements that make up modern warfare? Could any of these factors serve to more completely define modernity?

Modern armies are organized into army groups, field armies, corps, divisions, brigades, regiments, battalions, companies, platoons, and squads. Although generic field armies existed for some time, as did regiments and smaller units, divisions were created only in the latter half of the eighteenth century and corps at the beginning of the nineteenth century. The creation of divisions and, more specifically, corps altered the intellectual approach to the conduct of war. Armies divided only into units no larger than brigades had to be kept together; armies organized into divisions and corps could be dispersed across broad fronts and maneuvered according to a preconceived plan. This in turn lead to a different intellectual approach to the conduct of campaigns.

The Latin root of the word *campaign* means level countryside or field. In the past a campaign meant that armies would *go to the field* as opposed to being in garrison or winter quarters. A campaign did not mean a unique and grand military operation. The meaning of campaign changed with the breaking up of unitary armies into divisions and corps and their dispersal and maneuvers across and through a given geographic area. The separate divisions and corps were given specific missions that were part of a larger plan of *operations*. The deployment and use of different units in a theater of operations meant that there would be not one battle but a series of battles tied to a larger plan of operations or plan of campaign. Therefore, organizational change led to a change in the meaning of a campaign.

Leaving garrison and going to war during a specific season changed to a new type of military operation that not only firmly linked tactical engagements to strategic goals but created a new level of war. The deployment of divisions and corps and their maneuvers through a theater of operations is termed *distributed maneuver*. The creation of the link between theater-wide maneuvers and battle or battles became known as the *operational level of war*, positioned between the strategic and the tactical levels. *Operational art* is the process of actions and thought performed at this middle level. This is the process that determines military actions today. Thus, the rise of this new type of operational campaign can be one aspect that marks the beginning of modern war. Are there other factors to consider?

Distributed maneuver and operational campaigns required a new method of command. Army commanders could not be everywhere with corps moving and fighting along fronts over a hundred miles wide. Army commanders would have to rely fully on the initiative, judgment, effective communication, and tactical abilities of the subordinate commanders and their staffs. Broad mission orders would have to be issued to subordinate commanders, and the nature of command would have to be decentralized. For this system to work, both the army commander and the subordinate commanders and their staffs had to have a common understanding of tactical and operational methods, what is now called doctrine.

Although conscription began in antiquity, its use had all but disappeared until it was revived in the eighteenth century during the French Revolution. The reappearance of conscription enabled states to mobilize their manpower at a scale never seen before. Conscription began what could be called the mass age of military history. Military power based on the mobilization of the manpower and economic resources of the nation state, in particular conscript armies, became the norm in the nineteenth and twentieth centuries. With conscription, armies grew in size, which in turn led to the creation of larger formations such as army corps to more easily command and control them.

The other characteristic of armies of the nineteenth and twentieth centuries is their structural resiliency. The corps provided not only an effective structure for handling the large masses of men produced by conscription, but also served as the vehicle for nations in arms to recover from defeat and continue fighting. Corps were difficult to destroy in single engagements and could maneuver out of tight spots. Consequently, armies organized into corps often could not be destroyed in one battle but in a series of battles.

If one examines the patterns of operations in the last two centuries, one sees a pendulum swinging between apparent decisive victories on one end and long protracted affairs determined by attrition at the other. The relationship between the two is very often determined by the symmetrical or asymmetrical relationship between the opposing forces. Decisive battles and short wars occurred when the belligerents were out of balance or asymmetrical; protracted war resulted when the forces were in a rough equilibrium. The Napoleonic Wars, the American Civil War, and the two World Wars were protracted and strategically determined by attrition. Wars determined by attrition are nothing new and go back to antiquity. However, the clash between symmetrical armies has a major impact on our consideration of modern operational campaigns. When asymmetrical armies fight each other, single operations or decisive battles dominate military history. But when symmetrical armies fight, they are rarely overthrown in a single big battle; instead, victory in a campaign is a product of a series of related engagements. When rival armies both organized into corps fought each other, symmetry was achieved.

In addition to the development of the operational campaign within a theater of operations, modern warfare consists of the interrelationship of different campaigns in different theaters. Each theater of operations serves as part of the mosaic that forms a unified strategic war plan or grand strategy.

Let me propose a definition for modern war. A war is *modern* when it has all of the following characteristics: a strategic war plan that effectively integrates the various theaters of operations; the fullest mobilization of the resources of the state, which includes the raising of conscript armies; and the use of operational campaigns by opposing sides to achieve strategic objectives in the various theaters of operations. Those operational campaigns are characterized by symmetrical conscript armies organized into corps, maneuvered in a distributed fashion so that tactical engagements are sequenced and often simultaneous, command is decentralized, yet the commanders have a common understanding of operational methods. Victory is achieved by the cumulative effects of tactical engagements and operational campaigns. If one accepts these factors as providing the necessary conditions for modern war, then perhaps the first modern war was the one fought between Napoleonic France and Hapsburg Austria in 1809.

There have been some orthodoxies regarding the war of 1809 in the literature of Napoleonic history. It was considered that in this war Napoleon's powers of command started to decline. It has been stated that Napoleon made more mistakes in 1809 than in all his previous campaigns.[9] It may very well be true that Napoleon's personal effectiveness was decreasing by 1809. However, this interpretation comes from concentrating more on personalities engaging in war than an analysis of the process of war. The real story is that the Austrian army had reformed since 1805 and so the context of warfare had changed. If the context changed, then Napoleon's actions as a commander have to be reinterpreted. The story of 1809 is as much that of the Austrian army as it is that of the French.

Most accounts of this war concentrate on the main theater of operations in the Danube. The other theaters of operations, especially in Italy, are mentioned only in passing and are considered sideshows to the main event.[10] Therefore, the picture of Napoleon is incomplete, for he is examined more as an army commander than as a war strategist; this skews any effective examination of the war. This book will take a strategic as well as an operational view of the war as I examine the strategic missions assigned to the different theaters of operations and investigate how they collectively served to meet the strategic aims of the war.

By making both a strategic and an operational examination of the war, one is compelled to question another orthodoxy of Napoleonic history: the belief that Napoleon centralized all decision-making in his own hands and that he failed to train any of his subordinates in his methods of war.[11] I have always found this premise difficult to accept. I never considered it possible for one man to run a military establishment of a million men all by himself with all the rest waiting to be activated, especially with armies all over Europe. If this was the case, French armies would not have been as successful as they were. Napoleon's commanders had to have some understanding of what they were supposed to do. But one cannot speculate; there must be proof. So in this study, I intend to examine the relationship between Napoleon and his theater commanders, and in some instances his corps commanders, to see if he kept his methods of war hidden from his subordinates or if he tried to train them in the higher realms of the military art. If so, then one can say that French commanders had a common operational approach. This brings us to the issue of the modern concept of doctrine. The tactical regulations of 1791 served as a common drill book. But what about the broader concepts of doctrine as an accepted approach to the conduct of battles and campaigns? There were no formal, published doctrinal manuals above the tactical *règlements* ("regulations"). However, a comparative look at the operations of French armies in theaters where Napoleon was not in personal command may serve as a means to determine whether the broader concepts of Napoleon's operational methods were understood throughout the French army. If that proves to be the case, then one can say that Napoleon did indeed try to train his subordinates in his methods of war and the art of command. Furthermore, if there was such a dissemination of doctrine, for want of a better word, then the Napoleonic Wars take on further characteristics of what is referred to as modern war.

So what is this book about? It attempts to place the Franco-Austrian War in the context of the evolution of warfare in the nineteenth century. It will attempt to assess its significance. Was it just part of the Napoleonic Wars or did it signify a break in the pattern of the Napoleonic Wars? Did the nature of war in 1809 witness the emergence of unique characteristics that made it closely related to the wars later in the nineteenth century? Can it be said that the origins of modern war are found in 1809 rather than later in the nineteenth century? What does the history of this war say about the conduct of war in general? Did Napoleon change as a commander, or did the context of war change so that he looks different in comparison? How was the war fought strategically and operationally? What was the relationship between Napoleon and his commanders? Did he try to train them in his methods of war, or did he keep his methods secret, jealous of his power and fearful of rivals?

Finally, what I hope to convey is that history, especially the history of warfare, must be studied comprehensively—in length, breadth, and context. Any understanding of war must be gleaned from understanding that effectiveness in the conduct of war is always relational.

CHAPTER TWO

The Transformation of Warfare

The period 1763–1807 marks a significant transformation in the conduct of war. It witnessed such critical changes as to be considered revolutionary.[1] The period saw intellectual, political, social, and organizational changes in the conduct of war that led to the creation of a new type of army by 1805 and a new approach to war. Its use in a series of decisive campaigns in 1805–1807 marks the emergence of the first modern nineteenth century army.

In our own day, change in warfare is dominated by technology. This was not the case in the period under study. Eighteenth century weapons were, except for minor modifications, the same at the beginning of the century as at the end. Infantry weapons consisted for the most part of flintlock muskets or rifles and bayonets. Cannons were smoothbore muzzle loaders and cavalry fought with sword and lance. The critical change that occurred in warfare at the end of the eighteenth century was social, political, organizational, and intellectual rather than technological.

Army organization was the building block of the new warfare. At the start of the eighteenth century, the largest fixed military unit was the regiment, consisting of several battalions or squadrons. Higher units were organized on an ad hoc basis. Regiments were grouped into brigades and these were combined into larger formations designated as *columns* or as the *center, left,* and *right wings*. These formations in turn made up field armies. There were no permanent staffs above regimental level. Field army and formation staffs were personal members of a particular commander's entourage. The level of staff training varied. The division of field armies into *left, right,* and *center,* or *columns* was usually for temporary tactical purposes. This type of organization was cumbersome. Because of the absence of effective and permanent intermediate headquarters between regiments or brigades on the one hand and field armies on the other, commanders often had to concern themselves with the movements of individual regiments, which was time consuming and impeded anything beyond immediate tactical thought.

Field armies maneuvered and fought as unitary bodies. Each wing usually consisted wholly or primarily of cavalry while the center consisted primarily of infantry and artillery. Tactically, armies were often deployed in two lines or echelons of battalions or squadrons. Squadrons and battalions were usually deployed into battle lines several ranks deep, with artillery placed in the center. The main purpose of this deployment was so the infantry could deliver musket fire by volleys. The cumbersome command system combined with the necessity of deploying one's army into line often took considerable time. It also took a long time to redeploy. Consequently, if an army was caught in a poor position, as at Russbach and Leuthen where the Prussians were deployed against an enemy flank, it usually meant irreversible defeat. Therefore, initial deployment was critical.

The field armies of the eighteenth century usually averaged 50,000–60,000 men in strength.[2] Larger unitary field forces were usually beyond the capacity of the primitive system of organization and logistics. Clearly, a more sophisticated method of unit articulation was needed.

Warfare during this time was indecisive. Given the long time it took to deploy, big battles were usually by mutual consent. The ability to offer or decline battle and the size of these unitary field armies ensured that there was a dichotomy between the campaign and battle. It was possible to maneuver against an enemy for a whole campaign season without fighting a major engagement. Moreover, it was possible to win a campaign by forcing an opponent to withdraw solely by maneuver. It could be dangerous to risk losing a campaign by courting battle if one was winning a campaign by maneuver. And the opposite was true: a losing campaign could be reversed by the vagaries of battle.

Strategically, these armies were often deployed into different theaters of operations. The command structures and logistical limitations often made major strategic concentrations of several field armies difficult. The result was that a series of parallel campaigns fought *en cordon* were conducted by these armies. Each campaign was characterized by maneuver. If and when battles were fought, they often were not decisive. As a result, warfare in the eighteenth century was protracted, lasting many years—for example, the War of Spanish Succession, 1701–1713; the War of Austrian Succession, 1740–1748; and of course, the Seven Years' War, 1756–1763.

The creation of larger, more effective military formations took time. The French marshal Maurice de Saxe wrote of the creation of larger fixed formations in his *Reveries on the Art of War* in 1732. Saxe spoke of creating larger formations called *legions*, which consisted of four infantry regiments. Each regiment would have supporting units of cavalry and artillery and would total 3,579 men.[3] During the Seven Years' War, the French experimented with the creation of larger, more complex formations that could deploy into line of battle faster than before. This led to the formal creation of infantry and cavalry divisions in the army under Marshal Victor Broglie in 1763.[4] Each division was based on Saxe's model of four regiments; however, these regiments were larger than Saxe's. Two regiments were combined to form brigades while two brigades, with attached artillery, made up the division. Thus, the modern division was born.

The creation of divisions in the French army was part of a military renaissance that was going on throughout the French military establishment. Indeed, it was part of the larger intellectual movement that was the Enlightenment: an attempt to rationalize the world based on sound principles. This climate, combined with the poor showing of the French army in the Seven Years' War, gave further impetus to innovative thinking and experimentation. The most significant contributions of the military *philosophes* were those made by Pierre Bourcet and the Comte de Guibert.

Pierre Bourcet (1700–1780) was a staff officer in the French army and a principal adviser to a number of key generals. He was instrumental in the establishment of a staff school at Grenoble and while director of that school (1764–1771) wrote his *Principes de la guerre de montagnes* describing a new method of military operations. He realized that the newly created divisions could serve as semi-independent operational units beyond strict tactical use. The unitary field armies of the past could be broken into their separate divisions and dispersed on broad fronts. These divisions could then be maneuvered to turn enemy defenses as well as to concentrate against detachments, creating a force superiority at the decisive point. By moving forces along dispersed fronts, one could threaten multiple points and attack along different axes so the enemy would become confused. It would be more difficult to avoid battle since these divisions would be attacking from different directions. The method by which these divisions would alternately be dispersed and concentrated and dispersed again emphasized flexibility and initiative, which could create and exploit favorable tactical situations. This concept ended the dichotomy between campaign maneuver and battle and changed the nature of campaigns. Campaigns and battles became interrelated. In other words, by maneuvering to create favorable conditions for battle one was winning a campaign, and by conducting a favorable campaign one increased the odds that battles would be won. The chance of compromising a campaign by offering battle decreased. Bourcet thus viewed the relationship between campaign and battle as a continuous stream.

The intellectual approach to campaign planning had to be changed to accommodate this new structure and method. Bourcet wrote that a campaign plan had to have branches in order to adapt to the quick-moving, mobile operations that he theorized about.[5]

Bourcet also wrote that a campaign should be based on political objectives. That, of course, is nothing new. However, it is through this concept of linking strategic goals to campaign planning, and between campaign maneuver and battle or battles, that one finds the foundations of the modern concept of campaigns and of the operational level of war.[6]

If Bourcet dealt with the planning of campaigns, then the Comte de Guibert concentrated on the maneuver of armies and on the conduct of battle. An avid military thinker and writer, Guibert's most significant works were his *Essai general de tactique* (1772) and his *Defense du systeme de guerre moderne* (1779). In them, Guibert coined the term *grand tactics*, which he defined as "the science of generals,"[7] and included all of the "great parts of war."[8] Like Bourcet, Guibert wrote of the maneuver of armies consisting of divisions deployed at mutually supporting distances from each other. Most significant was his approach to the use of these forces in battle. Guibert believed that battles had a beginning, middle, and end. He believed that the beginning would be caused by the initial contact of an army's advance guard, a divisional formation that would engage the enemy force. The army commander would then maneuver his other divisions to arrive on the battlefield so as to mass to attack one sector while other divisions would defend in other sectors. The different divisions would attack from the march along separate axes. A climax would be reached as more and more divisions attacked. The enemy would be broken, to be followed by a pursuit.

This approach differed significantly from past practice in that the dispersed but mutually supporting divisions (what Guibert and later Napoleon referred to as being assembled) could be flexibly used to attack or defend depending on circumstances. The divisional system freed the army commander from concerns of minute tactical details. Since these divisions could attack from different directions, the likelihood of an enemy avoiding battle decreased. Armies could be pinned down by these detachments until the supporting divisions arrived to finish the job. This meant that armies would no longer move as unitary blocks. By moving in separate dispersed formations, the forces could be maneuvered to create favorable tactical situations. Because divisions could move on different axes and attack from different directions, the days of battles being fought only by mutual consent would end.

The successful use in battle of armies organized into divisions depended on the use of flexible tactics conducted by the units within the respective divisions. A debate over tactics grew within the French army. Should the French army copy the rigid linear formations of the Prussians? Should they rely on columns emphasizing the use of the bayonet? How should light infantry be used? Guibert's writings included a description of a flexible tactical system, based on the infantry battalion organized in a column of maneuver. The battalion would be trained to easily deploy into a linear formation and back into a column. The tactical commander would have the option either to deploy into a line for infantry fire or stay in column for a bayonet charge. Regimental commanders would have the option of deploying their battalions in mixed fashion of lines and columns known as *l'ordre mixe*. Guibert also wrote of cavalry and artillery tactics designed to enhance their combat effectiveness and stressed the interrelated use of the three combat arms. Reliance on the infantry column as a basis for maneuver and deployment enhanced the flexibility of tactical actions, allowing tacticians to overcome the limitations that terrain had placed on strict linear formations. This method served to complement the grand tactical objectives of the divisions and the field army.

Guibert's writings strongly influenced the tactical manual for the French army, the *Ordinance of 1791*. This document remained in force in the French army throughout the wars of 1792–1815.

Although not an operational issue, changes in the manufacture and use of artillery would have a future tactical effect. During the last quarter of the eighteenth century, artillery guns were becoming lighter in weight. Beginning in 1774, French artillery was standardized in respect to weight and carriage. Consequently, guns could be moved faster and easier. Ammunition was prepackaged to increase the rapidity and efficiency of fire. These reforms were conducted under the direction of Jean Baptiste Vacquette de Gribeauval and became known as the Gribeauval System. Doctrinally, the Chevalier Jean du Teil in his *De L'usage de l'artillerie nouvelle dans la guerre de campaign* (1778) stressed that the artillery should open the battle and be massed for decisive effect.

The result of the ferment and reform within the French military was that by 1791 the French army had an extremely flexible tactical system and was intellectually moving toward a distinct break with the older methods of waging battles and campaigns. The French Revolution had a further impact on the transformation of warfare.

What became known as Napoleonic warfare would have eventually emerged with or without the advent of the French Revolution. However, the Revolution speeded up the process. The French Revolution began in 1789 and by 1792, France was at war with most of Europe. The French army and society were initially disrupted by the Revolution. Many officers, particularly in the cavalry and infantry branches, fled or were killed. The country was invaded, civil war had erupted, and voluntary enlistments could not meet the emergency. Faced with military disaster, the revolutionary government mobilized the full manpower and economic resources of the state. The government reasoned that the French people were no longer subjects of the king but citizens of a republic. With expanded rights came further obligations to the state. With the old feudal social system swept away, the republic could exploit the state to an extent that no monarch could or would dare to do.

On August 23, 1793, the revolutionary government issued the *levée en masse*, decreeing that the entire French population was in permanent requisition for service in the war. All single men between eighteen and twenty-five years of age were to join the army. Married men were to work in the factories. Women were to go to the hospitals to help the wounded. Old men were to go to the public squares to exhort the locals. This was the first mobilization for total war in the modern era. The *levée* worked. By the autumn of 1794, over a million men had been mobilized, of which 850,770 were available for combat.[9] By 1794, workshops in Paris were producing 750 muskets a day, which almost matched musket production throughout the rest of Europe.[10] The *levée* was viewed as an emergency measure; national conscription was not regularized until 1798. However, the ability of the state to fully mobilize its resources is another block in the edifice of modern war.

The *levée* gave France the manpower and economic resources to wage war against the bulk of Europe. The army was flooded with untrained recruits, and to a great extent, the recruits were unable to execute the more intricate maneuvers of the 1791 field regulations. These enthusiastic but untrained troops fought in battalion columns or were dispersed into great clouds of skirmishers.[11] What occurred during the battles in the 1790s was a blending of the practice of deploying whole battalions and even regiments as light infantry fighting in open order and the flexible use of battalion columns for maneuver from the regulations of 1791. The result was a movement to an all purpose infantry that could fight like either light or line troops and a method of small unit tactics that was even more flexible than those laid down in the regulations of 1791.

A significant psychological change in methods of leadership occurred as a product of the Revolution. In the past, soldiers in the armies of the *ancien régime* were viewed as social inferiors of the officer class and had to be driven rather than led. However, the new French armies now consisted of citizen soldiers who were equal in social rank to the officers and refused to be brutalized; they had to be led by example. Moreover, the disruption of the officer corps because of the Revolution as well as the exigencies of war itself created a demand for good officers, many of whom rose through the ranks. The drastic situation placed a premium on leadership. Those who rose had to prove themselves to the government and to their men. Promotion in the armies became based on merit. Merit meant bravery under fire, showing initiative, and leading by example. Napoleon Bonaparte, as well as his future marshals Ney, Massena, and Lannes, catapulted to the top in a relatively short time. The promotion of such men to the highest positions set a tone within the army and established a standard for all to emulate. The passion for bravery and leadership permeated the entire officer corps and down to the lowest levels. It was for this reason that the armies of the Republic and later of the Empire fought with a particular daring and emotion that became known as *élan*. This spirit of initiative and leadership laid the foundations for a philosophy of decentralized command and control that would enable the Napoleonic armies to conduct mobile operations over broad fronts.

The new armies of the 1790s were organized and directed by Lazare Carnot, who headed the military section of the Committee of Public Safety. Carnot organized the field armies, which were in turn organized in divisions. Divisions were sometimes grouped experimentally into corps. He championed the rising array of talent in the officer corps and eventually assumed the operational direction of the war. The direction of Carnot, the new leadership, and the new tactics enabled the republic to survive and wage war, but it did not give it a decisive advantage.

In many ways, the War of the First Coalition, 1792–1797, resembled the previous wars of the eighteenth century. Strategically, the armies were deployed very much in cordon fashion with major field armies allocated for the Low Countries, the Rhine, Italy, and the Pyrenees. The armies were not evenly divided. The bulk of the troops were sent to the Low Countries and the Rhine. However, there was no overwhelming concentration of forces along a single line of operations. Carnot, like his opponents, still thought strategically in linear terms.

The rival forces were evenly matched. French victories were won at Jemappes, Hondschoote, Wattignies, and Fleurus. But there were victories for the opposing Allied coalition at Neerwinden, Amberg, and Würtzburg. Like other wars of the eighteenth century, this one was a long, drawn-out affair lasting six years. The Allied coalition was unable to concentrate all of its resources against France. Both Austria and Prussia kept substantial forces in eastern Europe, competing with each other and with Russia over the partition of Poland. Arguments over Poland finally induced Prussia to leave the alliance and make a separate peace with France in 1795. Spain also tired of the war and made its peace with France that same year. In 1796, Spain joined France in war against Britain.

France was able to maintain itself at full surge only during the years of the Jacobin Terror, 1793–1794. The successor government to the Jacobins, the Directory, was not as ruthless in mobilizing the resources of the French state. Moreover, French society was growing weary of the war and the economy was approaching collapse. Consequently, the strength of the rival forces declined by 1796, and a measure of strategic equilibrium remained.

The changes wrought by the Revolution did not ensure decisive tactical victories on the battlefield. The armies of the Republic were not effectively balanced among the three combat arms of infantry, cavalry, and artillery. The huge influx of recruits raised by the *levée* went mostly to the infantry. It took longer to train a gunner or trooper, and cavalry and artillery were more expensive than infantry. The greatest loss of officers during the early years was among the cavalry officers who were most representative of the nobility of the *ancien régime*. There was a lot of artillery in France, but the guns were parceled out in the 1790s among forces of the interior and within the many field formations on the frontiers. The cavalry of the Allied forces was superior at this time to the French and served to check the French skirmish tactics. Because the numbers of French artillery on the battlefield declined, the rival guns tactically tended to cancel each other out. Well-disciplined and well-led Allied infantry could effectively engage French infantry. As a result, there was a rough balance, tactically and strategically, between the competing forces. The changes in the French army had not given it a decisive edge over its opponents.

After years of uninterrupted warfare, Austria approached exhaustion and signed a peace treaty at Campo Formio in 1797 ending the war on the continent. However, the war between Britain and France continued. Napoleon Bonaparte took a military expedition to Egypt in 1798 in an attempt to establish a base for operations against the British in India. Bonaparte landed in Egypt and was stranded there after Admiral Horatio Nelson sailed into Aboukir Bay with a battle squadron and sank most of the French fleet. This action precipitated a declaration of war by the Ottoman Empire against France. Austria and Russia soon joined Britain and the Ottomans in a second coalition against France, bringing on another continental war.

French conscription had now become formalized into an annual levy. With the regular intake of recruits, strategic planning became easier. The campaign of 1799 was another seesaw affair with Allied victories in Italy and Germany, then a French victory near Zurich. The result was a stalemated situation by the close of 1799. Russia withdrew from the coalition leaving Britain and Austria as the only great powers at war with France. In 1799 Bonaparte returned to France after abandoning his army in Egypt. France was exhausted after a decade of revolution and war. The Directory, threatened internally from the political left and right, had already become dependent on the armies to remain in power, and the Directors were intriguing among themselves. Several believed that a more effective government needed to be created and sought to use a military man, Napoleon Bonaparte, to overthrow the government. The political plotters vastly underestimated their military paladin. The Directory was overthrown by coup d'etat in November. Thereafter Bonaparte preempted his political allies, establishing himself as the powerful executive, First Consul, of a new government known as the Consulate. For the first time, Bonaparte could combine his genius for war with the power of the French nation.

Some have argued that Napoleon was not an innovator but that he merely built upon the changes created by the Revolution: the full mobilization of the state starting with the *levée* and the tactical changes already made prior to and during the Revolution. It is argued that intellectually Napoleon relied on the thoughts of Bourcet, Guibert, and du Teil. In short, that Napoleon made but a practical application of all that had gone on before. This perhaps is not the entire case.

Napoleon's greatest contribution to the art of war was as a strategist, at least in the early years. Napoleon had a great ability to see the complete picture of war, to analyze all its different components, understand what was essential and what was not, and combine these factors into an integrated war plan and operational campaign plans. Napoleon was not the first to do this, nor would he be the last. Ulysses S. Grant, for example, had the same strategic grasp as Napoleon. What made Napoleon so unique was his ability to use the means that he had (tactical doctrine, conscript armies, etc.) and take them up to a higher level than had been done before or even imagined. Bourcet's and Guibert's writings about divisional based armies meant that there was now a clear relationship between campaign maneuver and battle. Napoleon carried this line further by conceiving of the strategically decisive campaign and battle. He sought to eliminate the enemy army in the major theater of operations in a single campaign: to create by campaign maneuver a tactical situation favorable to the French before battle began and to follow up with a battlefield victory so crushing as to render the state incapable of further resistance. Not only would the enemy be unable to fight another battle, it would be unable to continue the war. The enemy would sue for peace, the goal of the decisive strategic battle. Alexander the Great and Julius Caesar could win decisive victories, but they were unable to compel an unwilling enemy to fight at a disadvantage. Darius and Pompey, for example, had to willingly engage at Gaugamela and Pharsalus. The armies of the eighteenth century could avoid battle as well because there was no clear link between campaign maneuver and battle. Napoleon had the means to strengthen the link between campaign maneuver and battle; he conceived of armies and operations so large that enemy armies would be unable to avoid fighting. Through this connection, the operational campaign was created and the operational level of war was born. By this process Napoleon made the leap in warfare from the eighteenth to the nineteenth century.

The formation of the new warfare was based on the army corps, a multidivisional formation. Just as regiments and brigades were combined into divisions for greater striking power and easier command and control, the divisions would be combined into larger formations that were not only more powerful but provided greater resiliency in battle. The corps had been used experimentally in the French armies of the 1790s to ease command and control where there were a great many divisions, but these were not mature corps. Usually the senior division commander would command his own division as well as several others. This was the case, for example, with Massena's command in the Army of Italy in 1796.

The formal creation of army corps occurred on March 1, 1800, when Napoleon ordered General Jean Moreau, commander of the Army of the Rhine, to divide his army into four separate corps.[12] Each of the new corps was to consist of four infantry divisions, which would vary in size from 5,000 to 10,000 men. Consequently, the size of the corps varied from 20,000 to 40,000 depending on the size of the divisions. The cavalry of the Army of the Rhine was organized into divisions of 2,000 to 3,000 each. One cavalry division was assigned to each corps. One artillery battery of six guns was attached to each of the small divisions while the larger divisions got twelve guns. The army had a small artillery reserve. There were no artillery attached to the corps headquarters.[13] Napoleon also ordered the creation of an Army of Reserve that was to consist of three infantry corps and a weak cavalry corps. The corps and divisions were all staffed and included members from the general staff.[14]

Decisive action was what Bonaparte needed. Part of his political program and justification for seizing power was to bring order at home and peace abroad. Peace meant a reestablishment of the frontiers of 1797, the natural frontiers of France, including the Rhine, the Low Countries, and north central Italy.

The enemy armies totaled 108,500 on the upper Rhine and Danube under General Kray and an additional 100,000 commanded by General Melas scattered about northern Italy.[15] The main field forces of the French consisted of Moreau's Army of the Rhine with 104,000 men, Massena's Army of Italy with 36,000 troops, and the newly organized Army of Reserve concentrating in the interior at Dijon with about 45,000 in strength.[16] There were at least an additional 100,000 French troops in garrison and frontier duties.[17]

Bonaparte clearly saw that Germany was the major theater of operations and that the main strategic target was the Austrian army positioned between the upper Rhine and upper Danube rivers. The destruction of that army could knock Austria out of the war and end the conflict on the continent if not the entire war. This is what Bonaparte sought to do; this was the strategic objective.

The operational means to the strategic end was the rapid destruction, in a single campaign, of the Austrian army in Germany. What Bonaparte envisioned was something new. The French held Switzerland and could use it as a bastion to maneuver north into Germany or south into Italy. Bonaparte wanted Kray's army pinned down by a detached corps located in the Black Forest while the bulk of Moreau's army, supported by the Army of Reserve, would turn the Austrian left flank and come in behind them via Schaffhausen.[18] Kray's army would be trapped and forced to turn and fight the French on unfavorable ground to reopen their communications or be surrounded and destroyed. This great enveloping maneuver would become known as *la manoeuvre sur les derrières*. Once done, the Army of the Rhine would have the way clear for a drive down the Danube to Vienna, while the Army of Reserve could, if necessary, move through Switzerland to cut Melas' communications in Italy and destroy the Austrian army there in a second phase. This would be the decisive campaign whose centerpiece was the destruction of the main Austrian army in the main theater of operations—Kray's army.

However, this plan was not implemented. General Moreau considered the great enveloping maneuver too risky. Instead, he favored a conservative eighteenth-century approach by corps advancing abreast as a cordon with some emphasis to turn the Austrian left.[19] Bonaparte's political position was not yet strong enough to replace this recalcitrant general. Moreau was among the greatest generals of the French Republic, and the Army of the Rhine was as much his army as the Army of Italy had been Bonaparte's. Any attempt to remove Moreau could have precipitated a revolt in the Republic's largest and best army. In a meeting with Moreau's chief of staff, General Desolles, Bonaparte finally exclaimed in exasperation, "I shall carry out this plan, which he fails to understand, in another part of the theater of war. What he does not dare to do on the Rhine, I shall do over the Alps."[20] Bonaparte was compelled to abandon his great scheme to win the war and attempt his strategy of envelopment in a secondary theater of operations with far less means to achieve it.

Moreau began his conservative offensive, defeating the Austrians in battles at Stockach and Moskirch. Meanwhile, Bonaparte moved the Army of Reserve into Switzerland and on to Italy. The French emerged from the Alps and marched into the Lombard Plain, taking the Austrians completely by surprise. Bonaparte occupied Milan on June 2, effectively cutting the Austrian lines of communication north of the Po. However, Massena, who served as the pinning force in Italy by holding Genoa, was forced to evacuate the city under terms on June 5. The Austrians were now free to concentrate against Bonaparte and the Army of Reserve. Bonaparte moved east in early June looking for the Austrians. He mistakenly scattered his army hoping to secure all of the river crossings over the Bormida. The Austrians attacked Bonaparte's dispersed army at Marengo on June 14 and came close to routing the French. The timely arrival of Desaix's detached corps snatched victory from defeat. Although it was a close contest, the Austrians were beaten and cut off from their bases. An armistice was signed the following day in which the Austrians agreed to evacuate all of Lombardy. On June 19, Moreau inflicted another defeat on the Austrians at Hochstädt. A general armistice followed, but war resumed in November. That autumn Kray was replaced as Austrian commander in

Germany by the eighteen-year-old and inexperienced Archduke John, brother of the Hapsburg emperor. An encounter battle was fought at Hohenlinden on December 3 between 70,000 Austrians under John and 60,000 French under Moreau.[21] This battle shattered the Austrian army, which sustained 20,000 casualties.[22] The morale of the Austrian forces had cracked and with it, the will of Emperor Francis to continue the war. A peace treaty was signed at Luneville in February 1801, ending the war on the continent. Great Britain, now fighting alone, finally decided to give peace a chance and signed a treaty with France at Amiens on March 27, 1802. The wars of the French Revolution were over.

The War of the Second Coalition was a protracted war like the first. The victories won by the Allied forces indicated that they could effectively engage the French. The creation of army corps in the French army in 1800 and the change in operational orientation provided by Bonaparte indicated a move toward a different type of warfare. However, neither Marengo nor Hohenlinden seemed definitive in pointing toward a new type of warfare. The French *corps d'armée* needed to mature. Marengo was such a near run thing that the Austrians could attribute their defeat to bad luck rather than to any systemic inferiority to the French. The same could be said about Hohenlinden. The reforms in the French army in 1800 marked a transitional stage in the creation of a modern nineteenth-century army. It would take time for the structural reforms to take hold and for Bonaparte to establish effective operational control of all of France's armies. It was only after this maturation process that a new type of army conducting a new type of warfare would emerge in 1805.

The Peace of Amiens left the French Republic preeminent in Europe. France had reached its greatest territorial extent, with its frontiers established on the Alps, the Pyrenees, and the Rhine. In addition, French satellite republics were established in the Netherlands (called the Batavian Republic), Switzerland (Helvetian Republic), and in northern Italy (numerous republics that were later consolidated into the Italian Republic). Great Britain had hoped that in return for agreeing to France's expanded frontiers, the French would agree to a commercial treaty favorable to Britain. Not only did this not occur, but Bonaparte's aggressive policies regarding Genoa, Switzerland, the Ottoman Empire, and Santo Domingo indicated that France's appetites were not satiated. Fear of Bonaparte's continuing ambition prompted Britain to declare war on France on May 16, 1803, beginning a series of conflicts known as the Napoleonic Wars. The struggle between the leviathan and the behemoth would last almost unabated until 1815.

Britain, with the world's largest fleet, and France, with the world's most powerful army, remained stalemated. Britain blockaded French ports, and Napoleon concentrated an army along the Channel coast to prepare for a cross-channel invasion of Britain. Bonaparte's self-elevation to Emperor of the French and King of Italy (hereafter, Napoleon I) in 1804 and other aggressive diplomatic moves in Europe induced Austria and Russia to join in a coalition with Britain against France in 1805. Meanwhile, Spain had joined France in war against Britain. What followed was the War of the Third Coalition.

From 1803 to 1805, Napoleon concentrated a mighty host called the Army of England along the northern French coast. He had developed a naval campaign plan by which the combined French and Spanish fleets would lure away the British fleet to the West Indies and then double back to convoy the Army of England across the English Channel. Napoleon would land on the English coast, defeat the British forces with his overwhelming army, and dictate a favorable peace treaty from London. This was not to be.

Napoleon's scheme of luring the British battle fleet away from the Channel failed. The scheme drew off a portion of the British fleet under Nelson, but forces remained to cover the Channel. After a brush with Robert Calder's battle squadron off Ferrol, the combined French and Spanish fleet withdrew to Cadiz on August 20, where it was blockaded by Nelson's fleet. Meanwhile, Napoleon's eyes turned east.

The massing of the Austrian forces along France's and Italy's eastern frontiers compelled Napoleon to abandon the invasion of England. The French Emperor realized that he would have to strike and destroy the Austrian forces before the Russians joined them.

Austria's major territorial interests were in Italy. Seeking to regain Lombardy and smash French interests in Italy, Austria committed its largest field army of 100,000 there under Archduke Charles, its best commander. Charles' mission was to drive the French from Italy and threaten southern France; he would be supported by a joint Anglo-Russian amphibious operation that was to land in Naples. The Austrians believed that northern Italy would be the decisive theater of operations since that was where Napoleon had campaigned in 1796–1797 and in 1800. However, they were sadly mistaken; the Danube River valley that ran through the heart of the Austrian Empire was the decisive theater. Napoleon had been assigned to Italy in 1796 as a general of the Republic and was compelled to campaign there; he fought there again in 1800 for internal political, not military, reasons.

Austria committed a field force of 72,000 troops to the Danube under Archduke Ferdinand. This army would invade Bavaria, a French ally, as were the states of Württemberg and Baden, and await reinforcements from Russia. Two Russian field armies totaling over 70,000 would move from Russia through Austria via the Danube and unite with Ferdinand for a joint offensive across the Rhine. Additionally, Anglo-Russian amphibious operations were to be conducted against the Batavian Republic and Italy. A force of 22,000 men under Archduke John would serve to link the armies of Ferdinand and Charles by holding the Tyrol. In short, Allied and, in particular, Austrian strategic plans represented the cordon approach characteristic of previous wars. There was no decisive concentration of forces in the main theater of operations. There was no unifying operational theme but rather a scattered approach in which the different armies in separate theaters of operations fought parallel campaigns typical of eighteenth-century practice.

The Austrian army still represented the same old stand doing the same old business of eighteenth-century warfare. Although Austria had now lost two wars against the French, the losses were not decisive enough or traumatic enough to compel a major reform of the Austrian way of war. True, there had been battlefield defeats at Zurich, Marengo, and Hohenlinden, but there had been victories along the way as well. The view from Vienna was that in spite of the changes made by the French army in the 1790s, those changes did not give the French a decisive advantage on the battlefield. Some reforms had been attempted to improve the Austrian army during the years 1801–1805, but they were minimal and often stillborn. For example, Archduke Charles had been appointed president of the Hofkriegsrat (or Aulic Council) with instructions to reform the army. The Aulic Council had been the highest body charged with administrative duties of the armed forces and could advise the Emperor on military matters. It was highly bureaucratic and conservative and should not be confused with a modern war ministry.[23] Charles attempted to create a modern system of governmental administration, including a war and naval ministry, at the expense of the Aulic Council. He also attempted to put in a form of modern conscription, but this was blocked. Soldiers in the Austrian army still consisted primarily of long service professionals. Enlistment was set at ten, twelve, and fourteen years for infantry, cavalry, and artillery and technical branches, respectively.[24] A general staff for planning operations in peacetime was approved but remained only on paper until 1805. There was some improvement in administration, but that was about all that Charles could do. As war clouds threatened in 1803, Charles and his reforms waned, while the more conservative powers in Austria waxed. Emperor Francis was willing to listen to General Mack, who argued that further reforms were not necessary and that the fighting abilities of the army had improved, whereas Charles believed that this was not the case.[25] Emperor Francis was also fearful that his brother Charles would gain too much control of the military. And so, believing in "divide and rule," Francis in 1805 removed the Aulic Council from Charles' authority, restoring it to its old position of influence. Mack, Charles' opponent, was made Chief of the Quartermaster General Staff.

Tactically, the Austrian army still remained an eighteenth-century one. There were some reforms of infantry drill in 1804–1805; however, linear formations were still the norm.[26] What was worse was that there was no permanent fixed order for higher tactical formations above regimental level. Higher tactical units were organized into brigades and divisions on a semipermanent basis. Larger formations above the division were designated as *army detachments* and were organized in an ad hoc manner.[27] All of these units and formations were being hastily thrown together in 1805 as Austria prepared for war. These haphazard units and formations represented the usual eighteenth-century practice. The Austrian divisions and *army detachments* were not the equivalent of the French divisions and corps in that they lacked the cohesion of the more permanent French formations and missed the experienced staffs that could effectively coordinate the different combat arms. Although the Austrian military establishment had 1,800 pieces of artillery,[28] it lacked the requisite administration and structure to use them effectively in the coming war. The Austrian approach to war and the organization of the army was perhaps good enough to compete with the French of the 1790s, but the Austrians had failed to divine that Napoleon had forged an entirely new and devastatingly modern war machine.

The Russian army of 1805 was also still mired in the past. Some reforms had been attempted by Tsar Alexander I upon his ascension to the throne in 1801, but they were rudimentary at best. Staff work was primitive, and there were no fixed units or formations larger than that of the regiment. The creation of larger units called *columns* was totally ad hoc. With the low level of education found in the Russian officer corps and abominable staff work, the ability to effectively command large combined arms formations was practically nil.

The contrast between the Austrian and Russian armies and the French in 1805 is striking. The Army of England, soon to be renamed *La Grande Armée* (the Grand Army), was a new type of army and represented an advance from the armies of 1800–1801. It had been organized in the summer and autumn of 1804, and by August it consisted of 219,000 men and 396 guns.[29] But it was not the numbers and guns that made it unique; it was its structure and composition. The army was organized into seven corps, a cavalry corps, and a guard corps. The corps ranged in size from 14,000 men consisting of two infantry divisions to 40,000 men in four infantry divisions and a cavalry division.[30] On average, each corps had three infantry divisions, one cavalry division, and a small corps artillery reserve, and contained 25,000 men. The men of the army were citizen soldiers but included many veterans of the previous wars. The corps and division commanders were highly experienced but relatively young men. The corps commanders averaged thirty-six years of age. Each of the corps and divisions had established functioning staffs. A corps staff, for example, included a chief of staff, departments for intelligence and troop movements, several aide-de-camps, an artillery staff, an engineering staff including a half company of a bridging section, two companies of sappers and a company of miners, a logistical department, an ordnance unit, a gendarmerie unit,[31] and sometimes an ambulance detachment. It was both the structure of these divisions and corps and their familiarity with each other that gave them their particular cohesion and effectiveness. These forces had been together for almost a year. The commanders and staffs were working efficiently, and as a result, each corps and division commander could fight his command as an integrated combined arms unit. The years of peace had enabled the French to increase the artillery and cavalry components of their forces as well. Consequently, the corps of the French army of 1805 were more balanced, powerful, and effective than those of 1800 and 1801.

The high command of the *Grande Armée* was also unique. Napoleon combined the functions of chief of state, commander in chief of the armed forces, and commander in chief of the *Grande Armée*. Therefore, communication between Emperor and army commander was excellent. However, unlike the sovereigns of the other European armies, Napoleon was a real soldier and among the greatest commanders of all time. He had designed a staff system especially suited for his talents and objectives. Napoleon was the one who developed his operational plans rather than the staff. In this he was aided by his Cabinet, consisting of three bureaus. The Intelligence Bureau compiled enemy information. The Topographic Bureau maintained a situation map of enemy and friendly units. The third bureau, the Secretariat, wrote and dispatched Napoleon's orders. In addition, there was the Imperial General Staff directed by Marshal Alexandre Berthier. Berthier's staff was an executing rather than a planning staff in the modern sense. It oversaw the daily administration of the army, received reports from the corps headquarters, and sent information on enemy actions and the state of the French forces to Napoleon's personal headquarters. As constructed, it was the finest staff system of its day. As an instrument to execute Napoleon's operational designs, it was superb.

The collective effect of the staff system, corps structure, experienced citizen soldiers at all levels, and flexible tactics produced the first truly nineteenth-century army. The staff and corps system allowed Napoleon to effectively command and control a field army of over 200,000 for the first time in history. The separate corps themselves, with their own organic units, staffs, and flexible tactics, could fight a combined arms action with great advantage against an opponent of the same size but without the necessary structure, cohesion, and flexibility. This was clearly a revolutionary

army that outclassed anything else, especially if its operational use was as
dramatic as its form.

Unlike the Allies, Napoleon had a unifying operational theme to his
plan of campaign: to keep his enemies divided and decisively destroy
them in the critical theater of operations. Unlike Carnot, who developed
much of French strategy in the previous wars, and unlike his opponents as
well, Napoleon saw the conduct of war on the European continent holis-
tically rather than in segments as had been the usual procedure. In this
case, Napoleon had what was later referred to as *operational vision*. Ulysses S. Grant had the same gift. In 1800, Napoleon had already dreamed of
the destruction of the main Austrian army on the upper Danube by a single envelopment. Then he was thwarted by General Moreau, who had no
faith in the concept. Napoleon was compelled to execute a minor facsim-
ile in a secondary theater of operations—the Marengo campaign. Now Napoleon was in complete control. What he envisioned was a campaign
grander in scale than that of 1800.

Strength would not be frittered away into peripheral campaigns. There
would be just one major and one minor theater of operations. In the primary theater of operations, Napoleon envisioned a titanic sweep of an
army of over 200,000 men that would cross the Rhine and envelop the
Austrian army (Ferdinand's) from the north and destroy it before the Russians could arrive. The Russian detachments would be handled next,
opening the way to the heart of the Austrian Empire. Such a move would
hopefully knock Austria out of the war and lead to the collapse of the enemy coalition, if not peace. While Napoleon and the *Grande Armée*
moved against Ferdinand, Marshal Massena would take command of the
50,000 troops of the Army of Italy in the secondary theater of operations.
Massena was to engage Archduke Charles' powerful army and prevent it
from interfering on the Danube until it was too late. A corps of 20,000 under Gouvion Saint Cyr would move to cover any threat from Naples. A reserve army corps would hold the Batavian Republic, and the rest of
French coastal defense would be left to second- and third-line units. Nothing was being wasted.

Pierre Bourcet may indeed have provided the intellectual foundations
for Napoleon's education in the art of war. The sheer scope and size of
this operation surpasses anything Bourcet wrote about. It makes Napoleon, in this instance, an innovator in the art of war rather than a mere ex-
ecutor or compiler of the thoughts of the military *philosophes*. The total
of this campaign is greater than the sum of its parts.

The Army of England, now renamed *La Grande Armée*, was deployed
in August along the northern European coast from Brittany to Hanover.
Plans for the movement of this vast army from the coast to the Rhine and
then to the Danube were made well before the formal order to the army's

formations went out on August 27. In four weeks, 200,000 men had marched from the Channel coast to the west bank of the Rhine. By September 26, the *Grande Armée* was deployed along a 140-mile front from Neuf-Brisach to Mainz, fully equipped and supplied to begin the encircling operation. This four-week march from the coast to the Rhine was in itself an astounding piece of planning and staff work that shows Berthier's general staff at its best. Each corps had its designated march route. Supplies were assembled in advance of the march of the corps while in friendly territory. In some instances, soldiers, especially the Imperial Guard, rode in wagons rather than walked. A redeployment of so large an army in so short a time with such minimal wastage had never been done before. A century before it had been considered a great feat for the Duke of Marlborough to march 40,000 from the Low Countries to the Danube in the same time. In the 1790s, it was commonplace for Russian infantry regiments to sustain losses up to 50 percent on 200-mile marches due to wastage, desertion, and disease. The French had moved an army of 200,000 men a distance of 200 miles, bringing it into position intact.

On August 25, the *Grande Armée* crossed the Rhine on a 140-mile front. On September 2, Ferdinand's army invaded Bavaria, a French ally, and pushed on to Ulm with the intention of fortifying this crossing point over the Danube as a base. Ferdinand and his chief of staff, General Mack, expected the French to advance directly from Strasbourg via the Black Forest. Napoleon had Murat's cavalry corps and Lannes' V Corps create just such an impression as a feint. Meanwhile, the rest of the *Grande Armée* wheeled east and south, turned the Austrian right flank, and headed toward its rear.

The days of armies moving as unitary blocks under the direct command of the commander in chief were gone. It was impossible for an army commander to be everywhere and see everything along this moving 140-mile front. Napoleon understood this, and the corps structure and French staff system allowed the *Grande Armée* to conduct what later was termed *distributed maneuver*: the movement of major formations over a wide area according to a broadly conceived but flexible plan. Napoleon practiced what later became known as decentralized command and control. "Napoleon's orders were merely general directives; corps commanders were informed of the location and missions of the adjacent corps and told to maintain continuous lateral liaison."[32] On October 2, the *Grande Armée* approached the Danube in the Austrian rear. The next day, writing to Marshal Soult, Napoleon explained, "My intention, when we meet the foe, is to envelop him from all sides."[33]

The French crossing of the Danube began on October 6. The Austrian command was taken completely by surprise. Worse for them, neither Ferdinand nor Mack could fully comprehend just what was happening to them until it was far too late. The converging tentacles of Napoleon's army closed in on the Austrian forces. One Austrian detachment fled east and another ran south before the trap was sprung. As the French closed in on Ulm, another detachment with Archduke Ferdinand himself in command fled north; however, this column was driven into the ground by Murat's pursuing cavalry. The rest of the Austrian army, now commanded by Mack, was completely surrounded by the French in Ulm on October 15. The Russians were still over 100 miles to the east. On October 20, Mack surrendered. In the space of a four-week operation, an entire Austrian field army had ceased to exist, marking the triumph of a new kind of warfare by a new type of army.

With the Austrians disposed of, Napoleon turned east against the Russians. The Russian forces had been sent to Austria in two large detachments consisting of 40,000 troops under General Michael Kutuzov and 30,000 under General Wilhelm Buxhowden. By October 25, Kutuzov was about 100 miles east of Ulm, and Buxhowden was approaching Moravia. The *Grande Armée* hurtled east to devour Kutuzov. The Russians fled east, just managing to escape destruction. They retreated to Vienna, joining what few Austrian units they found along the way, and then turned north into Moravia hoping to link up with Buxhowden. The French were in hot pursuit. Vienna was taken on November 12. A final attempt to encircle Kutuzov just north of Vienna failed on November 15. Kutuzov linked up with Buxhowden and what remained of the Austrian forces in the area, giving a total of 85,000 for this conglomeration of troops. The forces of Archduke Charles were beginning to turn north from Italy and could threaten Napoleon's lines of communication.

Meanwhile, the battle power of Napoleon's army had declined. The French were compelled to leave forces behind to cover the flanks and rear of the penetration that ran along the Danube valley from Ulm to Vienna. After leaving detachments and garrisons, Napoleon had but 50,000–75,000 troops left for active operations unless he decided to withdraw. This would cancel out the gains of Ulm. Moreover, Napoleon learned that Prussia was preparing to enter the war on the Allied side. What Napoleon needed was one great victory that would shatter the coalition and knock one or both of his opponents out of the war.

Napoleon took his forces into Moravia seeking battle with the Allies near Olmutz. Anticipating a battle with a favorable numerical superiority, the Allies obliged. On December 2 at Austerlitz, the two forces met in battle. The combined Austro-Russian Army had 85,000 troops while the French had 75,000 on the battlefield. The Allied army's numerical superiority did not matter that much. The Allies were completely outclassed in respect to generalship, command, control, tactical flexibility, and intelligence. Discounting Napoleon's superior generalship at Austerlitz, the Allied army was an archaic organization. Units above the regimental level were hastily thrown together into larger columns. There was practically no effective staff system to coordinate the movements of the columns. Moreover, there had been no time to integrate any of the units to fight effectively as combined arms formations like the French corps.

An example of this mismatch is illustrated by the combat that occurred between the opposing left and right wings of the armies during the battle. The French left consisted of Lannes' V Corps and the Cavalry Reserve Corps under Murat. The Allied right wing consisted of Prince Peter Bagration's column and a cavalry force under Prince Lichtenstein. The opposing strengths were 20,100 French and twenty guns against 19,075 Allies with forty-two heavy and twenty-four light guns.[34] Notice that the numbers of troops were about even, but that the Allies had a distinct superiority in the number of guns. The two forces were to conduct a straightforward frontal battle; their respective missions were to engage and defeat the opponent before them. The French defeated the Allies in this sector and drove them from the field because the French forces fought an effective combined-arms battle that maximized the combat effectiveness of all arms. For example, Allied cavalry charged French infantry squares that broke their charge. The Allied cavalry in turn were countercharged by French horsemen while the Allied cavalry lacked support. After driving off the Allied cavalry, Lannes' and Murat's forces launched a combined arms assault against Bagration's infantry masses that drove them from the field.

In this instance, as in other sectors of the field, it was the superior ability of the French command and structural system that enabled them to maneuver forces and combine the effect of the combat arms. "What really happened was that we moved about a good deal, and that individual divisions fought successive actions in different parts of the field. This is what multiplied our forces throughout the day, and this is what the art of war is all about."[35]

Austerlitz was a decisive victory for the French and a disaster for the Allies. The Allied army was utterly shattered, suffering 27,000 casualties while the French lost 8,000.[36] The survivors of the Allied host were completely disorganized and incapable of further action or resistance, which was largely due to the absence of any effective structure above the regimental level. Austerlitz was the triumph of a modern army over an obsolete one; the contrast was clear.

Austerlitz knocked Austria out of the war and shattered the Allied coalition. With two armies destroyed in six weeks and his only remaining army, Charles', now hemmed in between the *Grande Armée* and the Army of Italy, Emperor Francis had no choice but to ask Napoleon for terms. A peace was signed later at Pressburg. Russia was incapable of further military operations and had also made tentative peace overtures to Napoleon. A European great power had suffered a defeat the likes of which had not been experienced in the eighteenth century. Napoleonic France had become the dominant power in central Europe.

Napoleon's victory did not sit well with Prussia. Napoleon knew that Prussia had been preparing to join his enemies before Austerlitz, and this treachery angered him. Now in the driver's seat, Napoleon compelled the Prussians to sign a treaty exchanging territory between the two powers. Prussia was given Hanover in exchange for Cleves and was forced to become a French ally and close her ports to British goods, precipitating a war with Britain. This policy damaged Prussia's economy and pride. The creation in 1806 of the Confederation of the Rhine, an alliance system between Napoleon and a number of German states, seemed to further threaten Prussia's interests. Finally, it was learned that Napoleon offered Britain the return of Hanover if London would make peace with him. This cavalier attitude toward Prussia's possessions drove the court at Berlin to war with France. The coalition was revived in July when Prussia and Russia signed a military alliance. Slowly, the Prussian army was mobilized for the coming fight.

The Prussian army was even more antiquated than the Austrian army. Save for the beginnings of a general staff under the reforms of Lecoq and Massenbach in 1800–1803,[37] the Prussian army slavishly followed the system of Frederick the Great. There was a strict adherence to linear tactics, and the army consisted of long-service professionals including a high proportion of mercenaries. Although the army had a considerable artillery train, the lack of unit articulation and effective command and control would render the artillery impotent in the coming campaign. It was only upon the outbreak of hostilities that the Prussians organized their forces into divisions.

In September 1806, the Prussians moved into Saxony, forcing that state to join them in war and incorporating the Saxon army with their own. Not wanting to wait for Russian support, the Prussians organized themselves into three field armies and moved west toward Hesse, seeking battle with the French.

Napoleon was happy to oblige. In a lightning campaign lasting less than four weeks, the armed forces of Prussia were completely destroyed. Moving at a speed the Prussians thought impossible, the *Grande Armée*, organized into three great operational columns, each in mutually supporting distance of the others, swept around the flank and into the rear of the Prussian armies. The Prussians turned, attempting to regain their communications, when their forces collided with the French near Jena on October 13. The following day witnessed the double battles of Jena and Auerstadt.

As October 14 dawned, both sides were still uncertain about the exact location of the opposing forces. The rival armies had been moving along a front roughly thirty miles wide. When contact was made around Jena, the Prussians believed that they were opposed by just a detachment, which was originally the case. Napoleon believed that the bulk of the Prussian army was to be found at Jena, when in reality it was merely the secondary Prussian army of Prince Hohenlohe with 35,000 troops. Assuming that he faced the main Prussian host, Napoleon planned a battle of annihilation by envelopment. The left and central columns of the *Grande Armée* (IV, V, VI, and VII Corps, Cavalry Reserve Corps, and Imperial Guard), totaling 90,000, were to pin the Prussians near Jena while the army's right wing of 50,000 troops (I and III Corps) were to envelop the enemy from the east and north. The forces of the right wing were to work in tandem. All did not go exactly according to plan. At Jena, Napoleon successively smashed the armies of Hohenlohe (35,000) and General Ruchel (13,000). However, the main Prussian army of 60,000 was being defeated ten miles to the north at Auerstadt by the 26,000 troops of the superb French III Corps. The main Prussian army, commanded by the Duke of Brunswick with the Prussian king, Frederick William III, was seeking a road to regain its communications when it collided with Marshal Davout's III Corps (26,000) at Auerstadt. The Prussian command became paralyzed when the Duke of Brunswick was killed. Davout was able to smash the piecemeal Prussian attacks, enfilade their battle line, and drive the Prussians from the field. Marshal Bernadotte, commanding the I Corps, was supposed to support Davout but failed to do so. The French were extremely lucky as the events at Auerstadt turned out.

The victories at Jena-Auerstadt were triumphs for the French system of war. The collision of large armies on extended fronts ensured that combat could no longer fall under the eyes of the respective commanders in chief. When operating in such extended fronts, mistakes (or in the modern parlance, *friction*) were bound to occur as they did in the misconception of reality in the minds of Napoleon and Frederick William and in the failure of Bernadotte to support Davout. Bernadotte's actions did not result from misunderstanding what his operational mission was but rather out of pique and jealousy. Bernadotte should have been removed from command because of his actions but was saved because he was Napoleon's brother-in-law. The corps system and Napoleon's decentralized system of command ensured that such mistakes during the confusion of operations did not prove fatal so long as the war was fought against an antiquated army.

Demoralized by defeats and lacking any structural cohesion, the Prussian armies disintegrated during the days following the battles of Jena-Auerstadt. On October 24, the *Grande Armée* marched into Berlin. By the first week of November, the French had taken 140,000 prisoners and 800 guns.[38] Save for a few detachments in Silesia and beyond the Oder River, the armed forces of Prussia had ceased to exist.

However, peace was not achieved. What was left of the Prussian forces moved into Poland and joined with the Russians. In ruthless pursuit, Napoleon advanced beyond the Vistula, conducting a winter campaign. The Russian forces had improved since Austerlitz. The respite enabled Alexander to raise a new army and to improve its structure by adopting the divisional system. Unlike the French, these divisions combined all three arms. Each division had eighteen battalions, twenty squadrons, and eighty-two guns. Initially, twelve divisions were raised; later the number was increased to eighteen.[39] Several divisions were grouped into field armies. (This was a step toward the creation of army corps in the Russian army, although corps actually were not created until 1812.) The objective of this reorganization was to enhance structural resiliency and to make command and control more effective, thereby improving the combat performance of the Russian army.

During the winter of 1806–1807, Napoleon tried and failed to trap the Russians as he had trapped enemy armies in the two previous campaigns. The Russians fought the French to a standstill at Eylau in February 1807. In that battle, the French came close to being defeated. The French VII Corps was practically annihilated advancing against a massed battery of artillery. The French could only claim victory because the Russians withdrew, leaving them in possession of a worthless battlefield. Eylau provided a preview of what was to come when the opposing forces improved the structure of their army. The intense firepower of the Russian guns was a by-product of the divisional system that improved the command and control of the Russian units. The ease with which the French defeated poorly articulated foes such as those at Austerlitz was gone. However, the Russian divisions were still not as effective as the French corps.

After a lull in which both sides went into winter quarters, the campaign resumed in March. Napoleon made several futile attempts to trap the Russian army under General Count von Benningsen. Finally, in June, Benningsen blundered into a bad position at Friedland, and Napoleon smashed his army. Benningsen had sought to pounce on what he believed was an isolated French corps. However, the operational flexibility of the corps system allowed Napoleon to rapidly direct reinforcements to Friedland. In the space of twelve hours, the French had concentrated 80,000 troops against Benningsen's 60,000. The superior combined-arms tactics of the French corps ensured another great victory. The French sustained 10,500 casualties while the Russians lost at least 18,000. Shortly after the battle, Tsar Alexander sued for peace. Friedland matched Austerlitz in its political effect and was in a similar league regarding the destruction of armies. However, the triumph masked some important trends. Because of the reforms in the structure of the Russian army, it was becoming harder for the French to outmaneuver and destroy a foe. The Russian divisional system was an improvement over the old column system but did not yet match the French corps system. What would happen if an enemy's organization and system matched the French was a question no one asked, let alone answered.

The treaty of peace was signed at Tilsit in July 1807. Russia ceded the Ionian Islands and Cattaro to France. Russia also agreed to offer mediation in favor of France in her continuing war against Britain. If Britain did not accept Russia's mediation, Russia would become an ally of France and join the Continental System (a French-led boycott of British goods). Napoleon, in turn, would support Russia in its war against the Ottoman Empire. Russia was to be given a free hand regarding Finland, and there was talk of a Franco-Russian partition of both the Ottoman Empire and British possessions in India.

At Tilsit, Prussia paid dearly for its losses and was compelled to join the Continental System against Britain. In addition, Prussia lost all of its territory west of the Elbe, which was turned into the Kingdom of Westphalia for Napoleon's youngest brother, Jerome. Prussia also lost the bulk of its Polish territory, which became the Grand Duchy of Warsaw and was given to the King of Saxony. Saxony and Westphalia joined Bavaria, Baden, Württemberg, Hesse, and other German states in Napoleon's Confederation of the Rhine, or Rheinbund. Prussia was compelled to pay an indemnity, and its army was limited to 42,000 men.

Tilsit represented the apogee of Napoleon's power, the fruits of the devastating campaigns of 1805–1807. Directly or indirectly, Napoleon ruled France, Italy, all of Europe between the Rhine and the Elbe, and the Duchy of Warsaw. Prussia was humbled and Austria beaten and isolated. Russia was now an ally. Only Britain remained at war with France.

The accumulation of such dazzling conquests was the result of the Napoleonic blitzkrieg. By 1805, Napoleon had created the first modern nineteenth-century army. The victories of Ulm, Austerlitz, and Jena were victories over obsolete armies that were still organized and trained as if for the eighteenth century. The Russians began to modernize in 1806 but were still severely outclassed, their divisional system more fitted to warfare in the 1790s. However, the reforms in the Russian army, as discussed earlier, indicated that the qualitative imbalance between the French forces and their opponents could be closed. The clash between modernity and the arcane produced a type of warfare in 1805–1807 that was short and decisive. If modernization spread to other armies, the nature of warfare would change again, and that is what happened in 1809.

CHAPTER THREE

Armies for Germany

Europe had changed since 1805. In that year, there were four continental great powers: France, Prussia, Austria, and Russia. The southern German states of Bavaria, Württemberg, and Baden were French allies. Francis of Austria held lands in Italy and could still claim to be titular head of Germany as Holy Roman Emperor. By 1808, the map of Europe was significantly different because of Napoleon's victories over Austria, Russia, and Prussia. Prussia had been badly beaten and humiliated in 1806. The 1807 Peace of Tilsit left it with only half of its former territory, having lost all of its land west of the Elbe and the province of Posen in the east. A huge war indemnity owed to France and the limitation of the army to 42,000 troops lowered Prussia to the status of a French client state. Russia, although beaten by Napoleon, had lost no territory and was now Napoleon's ally and still a European great power.

Was Austria still a great power? It had been beaten by Napoleon in 1805, losing Venetia, Istria, and Dalmatia to Napoleon and the Tyrol to Bavaria, and it had to pay a war indemnity as well. Besides those losses, there was a major shift in power relations, as Austria became ringed by Napoleonic power. The Holy Roman Empire was gone, and all of Germany between the Rhine and the Elbe consisted of French satellite states as members of the Confederation of the Rhine. On Austria's northwestern frontier was Bavaria, a French satellite state. The Prussian province of Posen had been turned into the Duchy of Warsaw, another French satellite, that bordered Austria from the northeast. The Napoleonic Kingdom of Italy and the French-held province of Istria threatened Austria from the south. Russia, now a French ally, posed a threat to Austria's eastern frontier.

However, Austria could still be considered a European great power. In spite of its many defeats, Austria had sources of strength. Politically independent with access to the Adriatic via Trieste, Austria was the third most populous state in Europe. The core of the Danubian monarchy remained, consisting of the Archduchy of Austria and the Kingdoms of Bohemia and Hungary.

Could Austria maintain its independence and its status as a great power? There was no love lost between Napoleon and the House of Hapsburg. Austria had been France's major opponent since 1792. Napoleon was

33

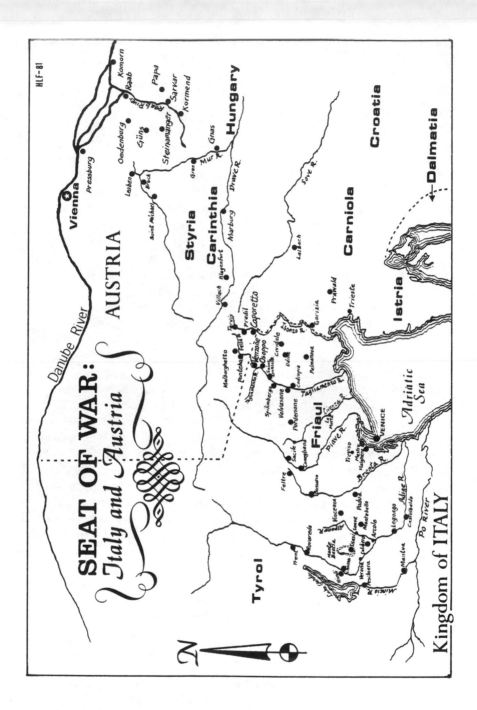

SEAT OF WAR:
Italy and Austria

viewed as a constant threat, and the Austrians realized that the French Emperor only respected power.

Austria was not safe as long as Napoleon ruled in Germany and Italy; the status quo was not acceptable. A restoration of Austrian power in Germany and Italy was necessary for survival. There would be another conflict with France; the only question was when it would occur.

Napoleon's eyes turned westward to the Iberian peninsula after the Peace of Tilsit. Britain was the last major power at war with France. Spain was a French ally, and Portugal was Britain's only remaining ally in Europe. The Continental System, Napoleon's strategic policy of depriving Britain of its European markets, was fully developed by 1806. Napoleon's original objective in 1807 was to subjugate Portugal in alliance with Spain. In November 1807, a joint French and Spanish force successfully invaded Portugal.

Napoleon also believed the Spanish monarchy to be corrupt and inefficient, which it was. The Spanish court was bitterly divided between two factions, one consisting of King Charles IV, his queen, and Prime Minister Godoy, the other headed by Crown Prince Ferdinand. Placing a French administration in Spain would ensure the effective direction of the Continental System there. Napoleon sent more French troops to Spain ostensibly to support the forces in Portugal. By May 1808, there were over 100,000 French troops in the peninsula. A Spanish uprising against Charles IV in favor of Ferdinand gave Napoleon his opportunity to replace the Spanish ruling house. Both sides requested Napoleon's mediation. The French Emperor forced both Charles and Ferdinand to renounce their rights to the Spanish throne, and he replaced them with his brother, Joseph.

The usurpation of the Spanish throne in the spring of 1808 was among the greatest strategic mistakes made by Napoleon. It showed the world that peace with him was impossible. The removal of the Spanish Bourbons precipitated a full-fledged insurrection in Spain in May 1808. Both the Spanish population and army rose in revolt. The French had never faced a nationalist insurrection on such a scale and were unaccustomed to dealing with guerrilla warfare. On July 20, a French corps that had been sent to pacify southern Spain surrendered to Spanish regular and partisan forces after it had been surrounded and cut off at Bailen. With the entire country in revolt, Joseph Bonaparte, the newly imported King of Spain, fled Madrid on August 1. Soon the French forces in Spain were retreating to the Ebro River.

The events in Iberia opened the way for Britain to land a field army on the peninsula. On August 1, a British army under Sir Arthur Wellesley, the future Duke of Wellington, landed in Portugal. On August 21, Wellesley defeated a French army under Junot at Vimerio. Wellesley was superseded in command by Generals Dalrymple and Burrand, who signed a convention at Cintra with Junot that led to the French evacuation of Portugal. The convention was unpopular in Britain because the terms stipulated that the Royal Navy was to return Junot's army safely back to France. Dalrymple and Burrand were soon dismissed. However, Britain could use Portugal as a base to intervene in Spain. By October, the new British commander, Sir John Moore, was advancing with an army from Portugal to support the Spanish forces. A new front had opened up against Napoleon in the west.

The war in Iberia offered Austria an opportunity: Vienna realized that the man who revolutionized warfare and carried political and social revolution into Germany and Italy could not be defeated by the old methods. Austria's armies had been obliterated in 1805 at Ulm and Austerlitz, and Austria needed a new type of army that could effectively compete with the French. The problem was that Napoleon's revolutionary army was a product of a revolutionary state. Could a conservative state produce a revolutionary army?

The very nature of the Hapsburg state precluded revolution, but would it prevent the creation of a modern army? The general consensus has been that armies represent their respective states. That is true, but does it mean that only nation states can produce effective armies with troops motivated enough to fight for the state? Imperial Rome and the British Empire produced efficient armies. Could Imperial Austria do the same?

Attempts were made to modernize the Austrian army directly after its defeat in 1805. Archduke Charles, Emperor Francis' brother and the only Austrian field commander to emerge with his prestige intact after 1805, was appointed Generalissimo of all Austrian armed forces in February 1806. Charles' mission was to get the Austrian army ready to fight the French.

Charles was a conservative in a reactionary state.[1] There were limits to what he could do personally and intellectually. He had an eighteenth-century mind rather than a nineteenth-century one. The intellectual approach of the previous century had favored the rationality, order, and centralization of the neoclassical age. Emotion and turbulence characterized the nineteenth century—the romantic age. Napoleon's intellect was more of the romantic than the neoclassical period. The style of warfare he practiced was violent, decentralized, confusing, and energetic, and his armies had that particular emotional quality, *élan*.

The Austrians were the first to be overwhelmed by the new warfare in 1805. The destruction of their forces at Ulm and Austerlitz was traumatic. The hallmarks of Napoleon's new warfare were speed, flexibility, and the large scale of operations. Based on the army corps system, flexible tactics, and a decentralized system of command and control, it was directed by a master of the art of war who understood that war was fast, violent, and confusing. The Austrian army of 1805, with rigid linear tactics, little unit articulation, and a penchant for centralized command and control, was totally outclassed; it had to change if it wanted to win.

Reforms were made in 1806–1809. A doctrine of war was developed among the Austrian officer corps by the publication of *The Fundamentals of the Higher Art of War for Generals of the Austrian Army*,[2] and other manuals on tactical doctrine were written for officers.[3] Attempts were made to make Austrian tactics more flexible by increasing the number of skirmishers who would fight in open order, but the doctrine still relied heavily on linear tactics.[4] However, it was an improvement over 1805.

Charles worked hard to increase and improve the artillery, and reforms were made to standardize the artillery system. Artillery transport was militarized, and the massing of the guns on the battlefield was stressed. When Austria created army corps in 1809, a corps artillery chief was included in each corps staff, which greatly aided in battlefield command and control of the guns. By 1809, the Austrian army had 742 field guns in 108 batteries.[5] These organizational improvements and the increase in the number of guns would eventually produce tactical firepower on a significantly grander scale than before. Reform of the artillery and its improvement was a technical matter and so was easier for a conservative state.

The French success had been in part due to strong unit cohesion created by humane treatment of troops and reliance on citizen/conscript soldiers. The Austrians tried to follow their example. To improve morale, new regulations were introduced to end corporal punishment, and the length of service was reduced. A limited form of conscription was introduced in the German lands of Austria and the Kingdom of Bohemia, and reserve forces were organized.

There was opposition to Charles within the high circles of the government. The Aulic Council or Hofkriegsrat, which in the past had served as the highest body for the direction of military policy, administration, and strategy, was made subordinate to the Archduke in 1806. However, this body remained and could block reforms or interfere with Charles' plans. Eventually, the Aulic Council regained its independent status to administer the army and provide strategic guidance.

The continued existence of the Aulic Council was a product of Emperor Francis' policy of divide and rule. In the final analysis, the Hapsburg army was the final guarantor of the monarchy. Hapsburg emperors were always fearful of the rise of another Wallenstein, who might usurp the throne. Totally loyal to his older brother and the concept of monarchy, Charles was unfairly suspected by Francis of having higher ambitions. Although recognizing the need for reform, Francis kept the Aulic Council to serve as a counter to Charles; the Hapsburg army would not become Charles' army.

The buildup of Austrian regular forces took time. Even among the reformers, conservatism was strong. There was some discussion of including the middle classes into the army and creating some form of militia or *Landwehr*. Originally, Charles and others in the Austrian high command opposed it. However, Napoleon's usurpation of the Spanish throne in 1808 heightened Austria's anxiety about its security. Austria felt compelled to create a *Landwehr* in the spring of 1808. The financial price of improving the army was high. It was cheaper to organize and train the *Landwehr*, which could augment the regular force by 180,000 men, rather than increase the regular army.[6] All males between 18 and 41 years of age in Austria and Bohemia were made liable for military service. For a conservative state, this was a significant step. The militarization of the population was viewed as being potentially dangerous to the existing social orders, but the removal of the Spanish ruling family by Napoleon had sent shock waves throughout Europe. Napoleon had usurped the throne of a friendly government; what would he do to Austria? The raising of a *Landwehr* was as an act of desperation to ensure the defense of the realm.

The Kingdom of Hungary opposed the military reforms. Hungary was the most feudal of the Hapsburg territories. The Hungarian Magyars expressed their will through the Hungarian Diet and resisted any attempts at Hapsburg centralization. The Magyars viewed the imposition of conscription or the raising of militia units as an infringement upon their traditional rights by the central government.

Here, Charles and even Emperor Francis were stymied. Any attempt to impose compliance would precipitate a revolt in Hungary led by the Magyars. The Hungarians agreed to allow Francis to mobilize the traditional feudal levy, the Hungarian *Insurrectio*, of 60,000 men in case of war and to provide 20,000 recruits for the regular army, but that was the extent of their cooperation. A further increase of Hapsburg strength would have to come from the hereditary and largely German-speaking territories of Austria and Bohemia.

Meanwhile, the French center of gravity shifted to Iberia. The defeats inflicted on French forces there meant that a further strategic commitment was needed. Units of the *Grande Armée* would have to go to Spain. The Spanish crisis produced a chain reaction in Europe: Napoleon's actions in Spain precipitated the Austrian call for a *Landwehr*, which in turn aggravated Napoleon's concern for his own security in central Europe. Napoleon realized that the movement of substantial forces to Spain would invite an Austrian move. To safeguard his rear, Napoleon sought the support of his new ally, Russia. A summit conference at Erfurt was held in October 1808.

Napoleon had yet to realize how badly the Spanish affair had hurt him. Prince Talleyrand, Napoleon's foreign minister, realized that Napoleon's ambitions were limitless and that Napoleon would eventually cause his own ruin. Talleyrand now secretly worked to save himself at the expense of his Emperor. Talleyrand was as cynical as Napoleon but could not be subverted by emotions as Napoleon was. Therefore, Talleyrand would prove to be the more realistic of the two. Napoleon needed Russian support at Erfurt. Talleyrand betrayed Napoleon by passing secret information to Tsar Alexander and advising him to stand firm against Napoleon.

On paper, the Erfurt summit seemed like Napoleon's triumph. Russia formally agreed "to make common cause with France"[7] in the event of Austria declaring war on France. The actual result was that the Tsar Alexander left the conference convinced that he could safely distance himself from his treaty commitments and the French alliance. By December 1808, the Russians communicated to the Austrians that they would make but a pretense of honoring the treaty.[8]

Napoleon, unaware of this, now based his central European strategy on the Russian alliance. Napoleon disbanded the *Grande Armée* and planned to take 125,000 of its troops with him to Spain. This left 120,000 troops in Germany, which were renamed the Army of the Rhine. The bulk of the army consisted of 80,000 troops of Davout's III Corps located west of the Elbe. There were other French detachments in garrison at Stettin, Pomerania, Danzig, and other areas in northern Germany. In addition, there were available 100,000 Allied troops of the Confederation of the Rhine.[9] Napoleon hoped these forces would keep the Austrians in line. Thus assured, Napoleon marched off to Spain.

Convinced there was an opportunity at hand, Austria strove to form a new coalition. Intrigues were begun with Russia and Prussia. In October, the Prussians secretly promised to join with Austria in war. The Prussians promised to commit 80,000 troops in a joint offensive against Napoleon along the Main River in central Germany. An alliance was made with Britain. The British promised financial subsidies to Austria of £750,000 of specie and £4 million in bills of exchange. In addition, Britain promised to send an expeditionary force to the north German Coast.

A meeting of the Crown Council was held in Vienna on February 12, 1809, and it was there that Austria made the final decision to go to war. The decision was based on a series of strategic assumptions: that Russia would remain neutral and that Prussia would join Austria and Britain. It was estimated that Napoleon, due to his commitments in Spain, would have only 200,000 troops for the German theater of operations while Austria alone could commit 300,000.[10] The strategic correlation of forces seemed strongly in Austria's favor. The original target date for war was March 1809.

Austrian planning and preparations continued. The result would be a surprisingly modern force (modern in a nineteenth-century sense). For the first and perhaps the only time, Austria tried to wage a nationalistic war. Great emphasis was placed on fostering German nationalism within the hereditary lands of Austria and Bohemia. Appeals were made for a nationalist rising against the French in other German states as well.[11]

None of the armies of 1809, however, could be considered to be ethnically pure. Napoleon's forces had a high proportion of German, Italian, and Polish troops in the "French" armies of 1809. Yet Napoleon appealed to the nationalism of his French-speaking forces as Francis appealed to his Germans.

The military reforms of 1806–1809 bore fruit. Austria was able to mobilize an army greater than ever before. The military forces of the Austrian Empire totaled 594,216 men, of which 283,401 were frontline troops, and included 742 field guns in 108 batteries.[12] The artillery complement was impressive. To top off the modernization attempts an army corps system was implemented for the field forces on February 2, 1809. There would be nine line and two reserve corps. Each line corps had twenty-five to thirty battalions, sixteen squadrons, and 64–84 guns. Each corps was divided into three divisions, one of which would serve as an advance guard division. The advance guard division consisted of a brigade of cavalry, a brigade of light infantry, and two light artillery batteries. The other divisions would be divided into three infantry brigades, each brigade of two regiments. Each brigade had its own artillery battery as did the cavalry brigade. Artillery batteries were sometimes attached to divisions and there was a corps artillery reserve of two to three batteries. The average strength of a corps ranged from 29,000 to 31,900.[13] The reserve corps were formed from heavy cavalry and elite grenadiers that were formed by drawing grenadier companies from the line regiments and combining them into battalions, brigades, and divisions.

Each corps was an effective combined arms unit capable of independent action. Although the corps staff were new and lacked the experience of their French counterparts, this was an improvement over the past, and it gave the Austrian army greater structural integrity and resiliency. The presence of corps artillery and staffs would ensure that the guns would be used effectively, permitting the Austrian corps to lay down a considerable amount of firepower. How to employ this imposing array?

The Austrian objective was to drive the French behind the Rhine and the Alps, destroying Napoleon's European empire and restoring Austrian primacy in Germany and Italy. This was an offensive strategy. There were to be three theaters of operations: Polish, Italian, and German. The Polish theater was the least important. Archduke Ferdinand, commanding the VII Corps in Galicia, was to invade the Duchy of Warsaw. The Austrian Army of Inner Austria under Archduke John, consisting of the VIII and IX Corps, would invade Italy, as well as Dalmatia, Istria, and the Tyrol. The

Main Imperial and Royal Army under Archduke Charles, consisting of two reserve corps and six line corps, would invade Germany. (The Italian theater is discussed in detail in later chapters.)

The German theater was clearly the primary one. Relying on the expected Prussian alliance, Archduke Charles would mount an offensive from Bohemia toward the valley of the Main River while a supporting offensive would be conducted along the Danube. By March, six of the Austrian corps were deployed in northwestern Bohemia, and two were on the Inn River south of the Danube. An offensive in March would have caught the French completely off balance. However, one of the strategic pillars in Austria's planning collapsed when Prussia changed course and refused to join the coalition. Austria's German strategy was dealt a severe blow, and the foundation for operations along the Main was destroyed. Charles still favored an offensive from Bohemia but was overruled by the Aulic Council, which favored a redeployment of the bulk of the army south of the Danube to mount an offensive into Bavaria. This would delay the offensive until April. Critics have argued that the delay gave Napoleon more time to prepare, and hence the entire campaign was compromised before it began. However, the reasons for the change were sound.

Without Prussia, a campaign on the Main was a blow in the air. An advance from Bohemia to the Main could be threatened in flank by a French offensive into Bohemia from Bavaria. On the other hand, an Austrian offensive up the Danube would protect the heart of Austria and place the main Austrian army in a better position to support an uprising in the Tyrol as well as cooperate with the Austrian campaign in Italy. As events would prove, the Italian and Danubian theaters were interrelated. The loss of the Prussian alliance unnerved Archduke Charles. He began to lose confidence in the possibility of victory. This created a pessimistic tone in all his operations that would eventually impede his effectiveness as a commander.

What was Napoleon doing? The Emperor moved into Spain with his army in the autumn of 1808. The regular Spanish forces, which were as antiquated in organization and tactics as Napoleon's previous opponents, were easily shattered along the Ebro River by Napoleon's offensive. Madrid was reoccupied by the French on December 3, and Napoleon reinstalled his brother Joseph as King of Spain. The Emperor then turned his attention to the British army operating in Spain under Sir John Moore. Napoleon went after the British with a vengeance, driving them into northern Spain. The British were heading for embarkation at Coroña when the Emperor decided to leave Spain.

Word reached him of Austria's preparation for war. By January 1, 1809, the news from central Europe was so threatening that Napoleon decided that he must go to Paris. Rumors of intrigues against him by Fouche (his chief of police) and Talleyrand did not help. The pursuit of Moore was left to Marshal Soult. The British army was safely evacuated, but Moore was killed in a rear-guard action. The war in Iberia was far from over. The British still held Portugal. Many regions of Spain remained unpacified, and the guerrilla war against the French continued. Napoleon would now have to prepare for a two-front war, although he still hoped that the Russians would keep Austria in check. Napoleon left Valladolid on January 17 and was back in Paris by January 24. Once central Europe was settled, Napoleon hoped he could finish with the Spanish business. This was not to be. Napoleon never returned, and the war continued there until 1814.

Napoleon had been considering the possibility of war with Austria since the spring of 1808. Contingency plans had been made, and preparations were under way for reinforcements for Germany and Italy. The Napoleonic empire had a great many resources to tap. In September 1808, 80,000 French conscripts were called up from the 1806–1809 classes. In December, an additional 80,000 were called from the class of 1810, and units in Spain were sent to reinforce Germany and Italy. In January 1809, Napoleon ordered the mobilization of an additional 110,000 from the class of 1810.

The German Allies, in particular the states of Bavaria, Württemberg, Baden, Saxony, and Westphalia, would be able to contribute at least 100,000. The Duchy of Warsaw could put 20,000 into the field. An additional 130,000–150,000 French troops would be sent later to Germany.

Most armies of the German states, like Austria, were also modernizing. Each of the members of the Rheinbund was required by treaty to provide a certain quota of troops for imperial defense. Bavaria, Württemberg, Baden, and Westphalia had all adopted the French system of flexible tactical drill and had adopted conscription and French methods of leadership as well. The armies of Bavaria, Württemberg, and Westphalia were organized into divisions. Similar organization and tactics helped in operating with French forces. The reforms greatly improved the military performance of German troops. In the forthcoming war, they would prove to be a match for the Austrians and in some cases, would fight as well as the French.[14]

This was not the case with the army of Saxony. Although a French ally, the Saxon King resisted modernization. Tactical drill and recruitment remained based on the outdated Prussian model. Serfdom continued in Saxony, and membership in the officer corps was determined by birth rather than by merit; the rank and file had to be cowed rather than led. The Saxons adhered to the strict linear tactics of the eighteenth century. Centralized routine was prized, and initiative was stifled. The Saxon troops would be combined into divisions, but these were more like the antiquated types of the eighteenth century. Consequently, the Saxon army remained an antique in 1809.[15]

On March 30, 1809, the French and Allied troops in Germany were designated as the Army of Germany. The French forces, as usual, were organized in army corps and were composed as follows: II Corps, commanded by General Nicholas Oudinot (14,000); III Corps under Marshal Louis Davout (45,000); and IV Corps under Marshal Andre Massena (40,000).[16] The French Imperial Guard and other units would arrive later. Marshal Alexandre Berthier served as chief of staff.

The Emperor also declared that the contingents of the larger Rheinbund states would be designated as corps and serve under French commanders. Some of the German monarchs protested but to no avail. The result was that internal administration and minor tactical deployment were left to the German commanders. Grand tactical and operational command of the corps was provided by the French commanders. Since the size of the Rheinbund forces were fixed by treaty, the size of these corps varied considerably as did their organization. The Bavarian forces sent to this army were designated as the VII Corps (30,000) and placed under the command of Marshal François Joseph Lefebvre. The Württemberg contingent became the VIII Corps (12,000) and was commanded by General Dominique Vandamme. Marshal Jean-Baptiste Bernadotte was given the Saxon contingent which became the IX Corps (19,000).[17]

The various corps commanders were in all cases equal or superior to the Austrians. Davout, Massena, and later Marshal Jean Lannes were, except for Napoleon, the greatest commanders in the French army. Lefebvre and Oudinot were excellent corps commanders. Bernadotte was perhaps the least able of the marshals. Although personally brave and a good tactician, he was difficult to work with. He had an inflated opinion of his abilities and tended to disdain his colleagues as well as the Emperor. He owed his position to his relationship by marriage to Napoleon. Vandamme, although a good tactician, did not get on well with the German troops under his command. Berthier, although perhaps out of his depth as an independent commander, was the best chief of staff in Europe, and the Imperial General Staff that he directed was the best executing staff in the world. The French corps commanders were more experienced at commanding and controlling army corps than the Austrians. The same can be said for the French division commanders and staffs.

French infantry divisions usually had only two artillery batteries per division and two batteries for a corps artillery reserve. There were no batteries at brigade level or at units below brigade at the time. The Austrians, by comparison, had a battery per brigade in each division and usually had three batteries in corps reserve.[18] French tactics were extremely flexible, and the French excelled at combined arms integration, a product of their greater experience, better staffs, and training.

Rheinbund divisions were organized a bit differently than the French.

Usually the divisions combined all three combat arms. The divisions would often have two infantry brigades, a cavalry brigade, and divisional artillery. The Rheinbund corps also had corps artillery reserves.[19]

What was Napoleon's strategy for the war and the campaign? Napoleon, almost up to the last minute, hoped that peace would be maintained. He wanted to give Austria no provocation. Austria would have to attack to trigger the Russian alliance. As such, French strategy would initially be to defend in all theaters. Only after the blow had been absorbed would offensive operations begin. Prince Poniatowski would command in the Duchy of Warsaw and Prince Eugene would command in Italy. Napoleon would direct all theaters as supreme commander and would take personal command of the Army of Germany once hostilities began.

Napoleon's initial plan for Germany was flexible and open-ended. The Emperor did not know if the Austrian offensive would be north of the Danube toward the Main River or south of the Danube into Bavaria. Therefore, he would plan for either contingency. Napoleon's army would be deployed in a great diamond formation, the points of the diamond being Ratisbon, Bamberg, Ulm, and Munich. Thus positioned, the army would meet frontally any attack along the Danube and would threaten the flank and rear of any Austrian offensive toward the Main.

This sketch of operations was spelled out to Berthier in a series of strategic instructions, the most important being on March 30, 1809. In this dispatch, Napoleon anticipated that hostilities would begin on April 15.[20] He explained to Berthier that if the Austrians moved toward Ratisbon, then the French would concentrate at Ingolstadt and Donauwörth and there engage the Austrians. Napoleon elaborated on further options depending on enemy actions: "What if the enemy moved on Nuremburg? He will find himself cut off from Bohemia. What if he moves towards Bamberg? He will equally be cut off. Finally, what if he marches on to Dresden [in Saxony]? Then we will enter Bohemia and pursue him in Germany."[21]

By occupying the above positions, Napoleon expected that an Austrian offensive on either bank of the Danube could be countered by an eventual French concentration followed by an encounter battle. If on the other hand, the Austrians developed an offensive toward the Main or Saxony, Napoleon would be in a position to turn their operational flank and rear. For any French offensive into Austria from Bavaria, Passau would serve as the French base. For this end, Passau was fortified and provisioned.[22]

A French offensive south of the Danube would be easiest since the roads were better on the south bank than on the north. At this stage, Napoleon's basic concept was to keep the components of his army within supporting distance of each other. A more detailed plan would have to wait until the Austrians revealed their hand. Napoleon planned to remain in Paris until the enemy attacked. Once the Austrian offensive began, Napoleon would go to Germany to take command. In the meantime, Berthier would be at Strasbourg. There he could be linked by manual telegraph to Napoleon in Paris. As chief of staff, Berthier would pass on orders from Napoleon to the corps commanders.

In early April, French intelligence indicated a shift of emphasis of the Austrians' main army to the south bank of the Danube. Napoleon indicated that if the Austrians attacked before April 15, the army should concentrate on Donauwörth. If they attacked on or after that date, the army would concentrate at Ratisbon.[23] The belief was that the French forces would be closer to the Austrian frontier and more prepared to meet the Austrians as the month progressed. Napoleon's plan was flexible and sketchy. Berthier was merely to serve as a cipher, and Napoleon fully expected to take command in the theater. Moreover, Napoleon's marshals were experienced and did not need detailed instructions.

Napoleon could not command in the other theaters of operations, such as in Italy. There, Prince Eugene de Beauharnais would be commander. Eugene was new to the art of command in war, and although Italy was the secondary theater of operations, it was strategically important. The German and Italian theaters would prove to be interdependent.

CHAPTER FOUR

Armies for Italy

The Italian peninsula was divided into three parts. The southern half constituted the Kingdom of Naples, ruled by Napoleon's brother-in-law Marshal Joachim Murat and his wife, Napoleon's sister, the treacherous Caroline. The Grand Duchy of Tuscany in east-central Italy was ruled by another Napoleonic sister and brother-in-law, Else and Prince Borghese. The final portion was the Kingdom of Italy, comprising north and west-central Italy. Rome at the time was occupied by the French but not yet formally annexed. The Province of Dalmatia on the Adriatic coast was also part of the Kingdom of Italy although it was separated by Austrian territory. Napoleon was King of Italy and ruled through his Viceroy, Prince Eugene de Beauharnais.

Eugene was one of the many interesting figures to take part in the Napoleonic saga. Born in 1781, the son of Josephine and Alexandre de Beauharnais, he had lived through the Reign of Terror of the Revolution, although his father was guillotined and his mother came close to execution as well. Penniless after the Reign of Terror, Josephine had to live by her wits and her charms, becoming a mistress to Paul Barras, one of the Directors. It was with the encouragement of Barras and because of her political connections that Napoleon married her in 1796. Eugene had always wanted to be a soldier and was appointed a second-lieutenant of hussars in 1797. Service in Italy as Bonaparte's aide-de-camp followed as did action in Egypt during Bonaparte's campaign there. Eugene fought at Marengo as a troop captain in the Consular Guard. Rapid promotion followed as Napoleon ascended the imperial throne. Eugene became Colonel General of the *Chasseurs-à-Chavel* (a unit of light cavalry) of the Imperial Guard, General of Brigade, and Arch-Chancellor of State. In 1805 he became Viceroy of Italy and was given command of the Army of Italy after Austerlitz. The next year he married Augusta-Amalia, Princess of Bavaria, and was made heir-presumptive to the Italian throne. The years 1805–1808 were spent in administering the kingdom and raising troops. Now there was the prospect of an active military command.

With many of Napoleon's top commanders either in Spain or needed in Germany, it was understood that Eugene would command in Italy. Napoleon had taken great pleasure and pains in coaching Eugene in the art of rule as Viceroy of Italy. He would do the same to teach his stepson the art of war.

Italy had been a battleground in every war with Austria. Napoleon hoped that the Austrians would repeat their mistake of 1805 by sending their largest field army to Italy. The Emperor considered the Danube the primary theater of operations in a war with Austria. The Army of Italy would have the mission of engaging a powerful Austrian army should the Austrians invade as in 1805, or engaging the Austrian forces along their southern frontiers should Austria remain passive in the south. In either case, Napoleon viewed the Army of Italy as a means to support the main French effort in Germany.

Napoleon grew concerned about the defense of Italy as he learned of Austria's threatening moves while he was in Spain. Prior to the Erfurt summit, he had sent orders to Eugene for the construction of a fortified line along the Piave River to defend Italy in the event of an attack.[1] Napoleon envisioned a series of strong points along the river that would serve to screen the French concentrations and as bases to attack enemy detachments. The strong points would conserve French strength while forcing the enemy to spend a great amount of resources to overcome them.[2]

While Napoleon campaigned in Spain, Eugene looked to the construction of the river defenses and observed the Austrians. The Viceroy had excellent sources of information and duly reported Austria's growing strength to his Emperor. Eugene reported on the raising and training of *Landwehr*, the expansion of bases at Villach and Laibach, and the opening of Trieste to British goods and officers.[3]

Eventually, Napoleon was led to believe that he had to return from Spain to prepare for a new war with Austria. Napoleon ordered Eugene to increase the levy of conscripts within Italy, as he was doing in France. Eugene reported that the Piave would not be effective as a fortified barrier. The Emperor agreed with Eugene's observations concerning the defense along the Piave.

On January 13 and 14, 1809, while still in Spain, Napoleon sent a series of "Notes" to Eugene; these were operational plans of campaign for the Army of Italy in the event of war with Austria. One was for an offensive campaign, the other, defensive. Since they formed the basis for the campaign in Italy, they deserve to be examined in some length.

The fortresses of Palmanova and Osoppo, located between the Isonzo and Tagliamento rivers, were to play both offensive and defensive roles. Palmanova would serve as the base for an offensive eastward across the Isonzo into the Austrian province of Carniola, while troops based at Osoppo could secure the flank of such a move. Osoppo, located near the upper Tagliamento, could logistically support an offensive northward into Austrian Carinthia, while Palmanova could screen the French flank in this case. If the French were compelled to withdraw before superior forces, then Osoppo and Palmanova would hold garrisons that would compel the Austrians to leave greater forces to besiege them.[4]

Palmanova and Osoppo would form part of a broader defensive system. Napoleon's plan was that after the enemy left detachments to blockade Palmanova and Osoppo, they would be drawn against a fortified river barrier. Since the Piave was not acceptable, Napoleon chose the Adige River and its tributary the Alpone as the new defense line. The Adige was difficult to ford and already contained the fortress cities of Verona and Legnano along its banks. The Adige was to be linked to Venice by diverting the waters of the Brenta and so joining that fortress city to the Adige line. A series of fortified bridgeheads were to be built along the Adige line; dry areas were to be inundated to cover the bridgeheads to render the Adige impassable at many points.[5]

The next part of the "Note" discussed the operational use of the fortified line. Napoleon pointed out that should the enemy disperse to observe the various fortified bridgeheads, Eugene could use the river line as a screen to concentrate at one of the bridgeheads, rapidly cross it, and destroy the enemy detachments in detail. If the enemy chose to mass between Arcole and Legnago, then the French could move into Venice and sortie from there, cutting the Austrian line of communications.[6] Elaborating further, Napoleon pointed out that:

> One can only hope for a defensive line to have the following advantages: To render the position of the enemy so difficult that he throws himself into faulty operations, allowing himself to be beaten by inferior numbers; or, if the enemy general is a prudent engineer, he would be obliged to methodically carry the obstacles created at our leisure, allowing us to gain time. On the other hand, viewed from the French side, the fortified line can assist the [numerical] weakness of the French general by making his position so clear and easy that he cannot commit any great mistake, and finally it will allow him time to await support. In the art of war, as in mechanics, time is the great element between mass and power.[7]

In summing up these instructions, Napoleon concluded: "The more one reflects on this position the more one thinks that with 30,000 men one has nothing to fear from 60,000 enemy troops of equal value, or that at least we shall be able to gain several months time."[8]

Napoleon also sent a plan for offensive operations. The Emperor considered an Austrian attack likely and was more concerned with defensive rather than offensive operations. But once Eugene's army was built up, the Emperor considered a more active role for the Army of Italy. As had been mentioned above, the Province of Dalmatia was separated from the Kingdom of Italy by the Austrian territories of Istria and Carniola. General August Marmont held Dalmatia with his grandiloquently titled Army of Dalmatia; it was in reality a corps of two infantry divisions. Marmont was under Eugene's command. In the event of war and the offensive variant was tried, Napoleon envisioned an envelopment of the Austrian forces in Carniola by the forces of Eugene and Marmont. Marmont was to be based at Zara with 12,000 troops and was to strike north, drawing enemy forces away from Eugene. If checked, Marmont was to retire on Zara and await support from Eugene. Meanwhile, Eugene's army would strike due east from Palmanova. The two forces would converge at Laibach, destroying the Austrians between them. Once this phase was completed, the two armies were to move north to support Napoleon in Germany.[9]

Both defensive and offensive campaign plans for the Army of Italy incorporate the major themes of Napoleon's methods of operation. The defensive plan offered variations of the strategy of the central position by which superior forces would be brought to bear against detachments and the strategy of envelopment via the *manoeuvre sur les derrières* (see Appendix A). The offensive plan stressed envelopment by two independent forces, Eugene's and Marmont's, launched from two different provinces along separate axes. Napoleon did not seek to hide his methods of war. Both army commanders, Eugene and Marmont, were informed of the concept of operations and the role they were to play.

If one compares the instructions for Napoleon's commanders in Italy and in Germany, one is struck by the fact that Napoleon sought to explain his methods of war to his subordinate commanders. It was explained to Berthier in Germany, but both Davout and Massena seemed to understand the concept of operations as well.[10]

The southern frontiers of the Austrian Empire were defended by the Army of Inner-Austria commanded by Archduke John. The recruiting and supporting areas for the army included the Austrian provinces of Carinthia, Styria, Carniola, and Croatia. John's previous military experience included commanding the Austrian army at Hohenlinden in 1800 and the Austrian forces in the Tyrol linking the Austrian armies in Germany and Italy in 1805. Although his direct military experience was not extensive, he was the right man for the job for political and strategic reasons. John was among the leaders within the imperial family favoring German nationalism. He had been in contact with Tyroleans who wished to be free from Bavarian control. His political actions as well as his experience in 1805 meant that he was the right man to support an insurrection in the Tyrol; consequently, the Tyrol was given to John as part of his theater of operations.

John's army consisted of the VIII and IX Corps. The former, commanded by Feldmarschall-Leutnant Jean, the Marquis de Chasteler-Corucelle, was located in Carinthia with bases at Klagenfurt and Villach. The IX Corps, commanded by Feldmarschall-Leutnant Ignatius Giulay, the Ban of Croatia, was in Carniola with its main base at Laibach. Including *Landwehr* and other reserves, this force totaled 76,390 troops and 148 guns.[11]

As in Germany, Austria's war plans for Italy were offensive. This was a grave mistake. An offensive along the Danube made sense since the Austrian army there had a numerical advantage and that was the decisive theater of operations. But since the Austrians planned to commit only two corps to the Italian theater, no such superiority existed. The Austrians would be outnumbered. In such a situation, it would be better for Austria to conserve her resources. With over 76,000 men, the superb defensive terrain provided by the Alps, and with bases at Laibach, Villach, and Klagenfurt, the Austrians could have ensured the integrity of their southern frontiers by remaining on the defensive. Instead, this army would be sacrificed in a desperate offensive gamble.

Austrian plans called for a broad, diffused offensive that would coincide with insurrections in the Tyrol and in Italy. Chasteler was to take 10,000 infantry and 370 cavalry from the VIII Corps and invade the Tyrol to spark an insurrection there. Albert Giulay would take over command of the VIII Corps. Generalmajor Andreas von Stoichewich would take 12,000 infantry and 150 cavalry from the IX Corps to invade Dalmatia. This essentially reduced the invasion force for Italy to at most twelve brigades, too small a force to conquer northern Italy. However, this is just what was expected. John was to take the balance of the VIII and IX Corps and invade Italy proper. The main body of the Austrian army was to move down the valley of the Isonzo towards Cividale, Udine, and the central Tagliamento. Flanking forces would move toward Osoppo and Palmanova. The French were to be surprised, and it was hoped that John could move fast enough to destroy Eugene's army in detail. The Austrians had also sent agents to initiate an uprising in the Italian cities that would disrupt Eugene's rear. If all this worked, it was hoped that John could overrun north-central Italy. The invasion plans were finalized on March 18.[12]

The Army of Italy had to be made ready for war. In 1808, the army had only 20,000 troops on its rolls. Napoleon had hoped to raise its strength to 60,000 by April 1, 1809, and to 90,000 by the end of that month. A portion of the conscripts raised in France were sent to the units in Italy. Conscription within the Kingdom of Italy swelled the ranks of the Italian regiments in Eugene's army. Troops that had been sent from their parent regiments in Italy to fight in Spain or Naples were returned.[13] By early April, the Emperor's expectations were surpassed. The frontline strength of the Army of Italy, excluding the Army of Dalmatia, totaled 81,103 with

116 guns.[14] Marmont had an additional 12,000 men in two divisions and an extensive artillery force totaling 78 guns.[15] The Army of Italy consisted of eight infantry divisions, three cavalry divisions, and a small mixed division of the Royal Guard. Combined with Marmont's command, Eugene's frontline strength totaled 93,000 men and 194 guns. The balance of forces in Italy favored the French. In addition, fortresses were being stockpiled with ammunition and supplies. Vehicles were being concentrated to provide transport for the army. Fortifications along the Adige were being built.

However, the Army of Italy was not yet fully ready for war. Although the number of troops had been increased satisfactorily, there was still a need for 140 line officers to fully staff the army. In addition, there were not enough guns to provide adequate division, corps, and army artillery reserves, nor had the army been organized into army corps.

The delay in creating corps can be blamed on both Eugene and Napoleon. The Viceroy had no experience with high command in wartime. His last real combat experience was as a troop commander at Marengo. So far, Eugene's tenure as army commander had been mostly administrative. Eugene's division commanders had far more experience than he in commanding large formations in combat. Marmont, for example, had commanded the II Corps in the *Grande Armée* of 1805 and was commanding troops in Dalmatia. General of Division Paul Grenier commanded a corps at Hohenlinden. Eugene believed that it would be difficult enough to command his veteran division commanders; it would be almost impossible to command if any of Napoleon's marshals were sent to Italy. The appointment of one or more marshals to Italy was synonymous with the creation of corps. Eugene wrote to Napoleon to forestall such a move:

> I desire that Your Majesty does not send here any of the marshals. I feel strong enough to manage the army in such a way that would please you. . . . I would prefer to deal directly with my divisional generals. I know them all and would work better with them. They are all very good and do not have the pretensions of the marshals.[16]

Napoleon agreed not to send any of his marshals (they were largely committed to Spain or Germany anyway). However, the Army of Italy was too large and cumbersome to command effectively without a corps organization. Moreover, Napoleon knew that corps gave field armies a particular resilience and enhanced combat power. Writing to Eugene on the necessity of combining divisions into corps, Napoleon pointed out that "a single division by itself is too weak and would soon be reduced to 6,000. But two divisions of 18,000 can go anywhere."[17]

The two wrote back and forth on how best to organize the army, whether the new formations would be called *wings* or *corps*, and who would command them.[18] Eventually, three men were nominated as corps commanders.[19] They were Generals of Division Macdonald, Grenier, and Baraguey d'Hilliers.

Etienne-Jacques-Joseph-Alexandre Macdonald had considerable experience. He had served in Italy from 1798 to 1800 and for a time had commanded the French Army of Naples. Macdonald had been in forced retirement for several years because of the enmity of Napoleon. Macdonald had been a friend of General Moreau and had strong republican leanings. An ill-advised love affair with Napoleon's sister, Pauline, alienated the Emperor even further. Macdonald's appointment came via the intercession of Empress Josephine and her daughter Hortense. Paul Grenier was among the most experienced of the generals in the Army of Italy. He had served in the Army of the Sambre-et-Meuse from 1793 to 1798 and became a general of division in 1794. He commanded a corps in the Army of Germany in 1800 and had been with the Army of Italy since 1806. Louis Baraguey d'Hilliers had a distinguished military career. Currently serving as the military governor of Venice, he had campaigned with Napoleon in Egypt and Italy, commanded a dragoon division in the *Grande Armée* in 1805, and briefly commanded a corps in Italy in 1806.

Napoleon took so long to act on the nominations that they were not confirmed until after war had started and so delayed the creation of army corps. This lack of haste is curious. Trusting to the Russian alliance, Napoleon had trouble believing the Austrians were really going to attack him. Even if they did, Napoleon was convinced that there was still plenty of time to meet the attack. Throughout most of March he believed that Austria would not attack until May 1, and only by March 30 did he consider that war was likely any time after April 15.[20] Consequently, there still seemed to be plenty of time to organize the army's high command.

The structure of the Army of Italy was thus incomplete at the start of the war; it was scattered as well. Napoleon dispersed the Army of Italy for both political and logistical reasons. As in Germany, Napoleon sought to avoid a provocation that might result from a concentration on Austria's frontiers. Napoleon ceded the opening move to Austria to blame it for starting the war and to trigger the Russian alliance. Furthermore, a premature concentration east of the Tagliamento would place a severe strain on the logistical service. Magazines had been established in the area, but these were built with an eye to supplying the army once operations began. If the army was kept concentrated east of the Tagliamento in peacetime, the supply bases in eastern Venetia would soon be depleted and would be unable to support the army during either an offensive or a defensive campaign.[21]

As of April 1, the Army of Italy had two infantry divisions and a light

cavalry brigade between the Isonzo and the Tagliamento. Another infantry division was behind the Piave at Treviso. The rest of the army was east
of the Adige with units extending to Milan and Tuscany.[22] Although the
French had a numerical superiority, the dispersal of Eugene's army and its
incomplete command structure offered an opportunity to the Austrians.
They would have some surprises for the French in Italy and Germany.

CHAPTER FIVE

War along the Danube

Berthier was in Strasbourg receiving Napoleon's messages from Paris and transmitting them to the corps commanders. However, the Army of Germany's dispositions were such that it was ready to move should Austria attack. Napoleon's forces along the Danube totaled 165,000 troops (more were en route) and 311 guns.[1] The French deployment was broad, extending a length of 160 miles, but was well positioned to intercept the Austrians.

Davout's III Corps, consisting of five infantry divisions, two cavalry divisions, and one cavalry brigade, covered the front from Bayreuth to Ratisbon. Echeloned behind Davout between Nordlingen and Elwangen was Vandamme's VIII Corps. South of the Danube was Lefebvre's VII Corps with three infantry divisions deployed from Straubing to Anzing. A cavalry screen was operating along the Inn River, which formed the frontier with Austria. Echeloned behind Lefebvre was Oudinot's II Corps with divisions at Pfaffenhofen and Augsburg. Massena's IV Corps of four infantry divisions and one light cavalry division was further east of Ulm. An additional 22,000 troops were assembling at Strasbourg.[2] This was a broad and dispersed formation capable of maneuvering north or south of the Danube as well as moving east or west. Each of the various components of the army was within supporting distance from the others.

The Austrian redeployment was nearing completion by the first week in April. Charles' army totaled 209,400 men with 500 guns.[3] The Austrian army was organized into three massive formations. The right wing, consisting of Count Bellegarde's I Corps and Count Kollowrat's II Corps, a total of 58,000, would advance out of Bohemia to operate north of the Danube. Its mission was to support the main offensive south of the river and threaten Napoleon's line of communication running from the Danube to Strasbourg. South of the Danube was the Austrian center and left. The center consisted of Prince Hohenzollern's III Corps, Prince Rosenberg's IV Corps, and Prince Lichtenstein's I Reserve Corps, totaling 66,000. This force was to advance from Scharding to Ratisbon. Archduke Charles traveled with this force. The left wing consisted of Archduke Louis' V Corps, Hiller's VI Corps, and Kienmayer's (or Kienmaier's) II Reserve Corps and totaled 61,000. This wing was to serve as flank guard and advance along the axis Branau-Neumarkt-Landshut. The entire army would advance on a

100-mile-long front—70 miles south of the Danube and 30 miles north of the Danube.

The Austrian forces outnumbered the French, but the numerical disparity would eventually close. The best chance for an Austrian success was to attack quickly and defeat the various components of the French army before they could unite. Such a plan placed a premium on speed and rapid command and control. The new corps organization should in theory expedite speed in maneuver; however, the command and staff were untried, which tended to slow the pace. If the tempo of operations could not be maintained, then the campaign would be lost.

The other weakness of the plan was that once the offensive began the northern wing (I and II Corps) would be isolated from the rest by the Danube. Communication with the rest of the army south of the river would depend on capturing bridges in enemy territory. If bridges were not captured, the Austrians would have to build their own. (Notice the contrast to Napoleon's plan, which ensured easy maneuverability and communication between the two banks.) This was the best that Austria could do, unless they abandoned going to war at all, and things had progressed too far to stop. The only way Austria could win was by waging an offensive war. There would have been a greater chance of success had Prussia joined the alliance, but with Prussia's withdrawal, all that Austria could do was attack and hope that a German nationalist uprising would help them topple Napoleon. The Austrians hoped the south German states and the Italians would defect. Plans were made to foster nationalist revolts in the Tyrol and in Italy.

In spite of the drawbacks, this was a very "modern" plan of campaign, with operational depth as well as breadth. Austria would commit the largest field army it had ever raised, organized on modern lines, and attack along a broad front. Appeals to the ideology of national liberation were made. Insurrections were to disrupt the strategic or operational rear of the enemy force. For the first time, the French were going to fight a modern nineteenth-century army.

There was no formal declaration of war. On April 9, 1809, letters were delivered to French commanders in Germany, Italy, and the Duchy of Warsaw notifying them that the Austrians were going to advance and "treat as enemies all who opposed them."[4] The war had begun.

The Austrian offensive took the French momentarily by surprise. The incomplete command structure of the Army of Germany brought it close to disaster. Berthier was not empowered to act as commander in chief, and Napoleon was still in Paris on April 9. Berthier excelled at executing operational orders but was unwilling to show initiative.

Napoleon had told Berthier that if the Austrians attacked on or after April 15, a concentration of the army should be at Ratisbon. If the Aus-

CAMPAIGN OF 1809

RATISBON PHASE

Situation 10 April 1809, and Austrian
Advance From Concentration Areas

SCALE OF MILES
10 0 10 20 30 40 50 60 70 80 90

BELLEGARDE
commanded both corps.

XXX
PONIATOWSKY (27,000)
POLAND ARCHD. FERDINAND
(40,400)

XXXX
ARCHD. CHARLES
(209,400 including 15,000
artillery and engineers

BOHEMIA

DRESDEN

PRAGUE

KLATTAU

STRAKONITZ
V
IV
III

BUDWEIS

MORAVIA

SCHWEIDNITZ

OLMUTZ

BRUNN

AUSTERLITZ

IR

LAA NIKOLSBURG
ZNAYM

KLATTAU

HORN SCHÖNGRABERN
MISTELBACH WULFERSDORF

ZWETTL MEISSAU
HOLLABRUNN

Veczay (?)
(16,000)

Dedowich (IV)
(8,000)

LOWER AUSTRIA

KREMS STOCKERAU
MAUTERN KORNEUBURG
KLOSTERNEUBURG WAGRAM (DEUTSCH WAGRAM)
VIENNA ASPERN

V III IR
(6,000) (25,000) (18,000)

V IIR
(25,000) (9,500)

Jellacic (X)
10,000)

EBELSBERG ENNS MAUTHAUSEN PÖCHLARN
KLEIN-MÜNCHEN MÜLK ST. PÖLTEN
GMAUTHAUSEN AMSTETTEN
WELS LAMBACH STEYR
VOCKLABRUCK

LAXENBURG FISCHAMEND
BRUCK-ON-THE-LEITHA

HAINBURG PRESSBURG

RITTSEE

KOMORN

Danube R.

PUCKERSDORF

CORPS COMMANDERS

AUSTRIA

XXX
IR LICHTENSTEIN
IV ROSENBERG
III HOHENZOLLERN
V ARCHD. LOUIS
VI HILLER
IIR KIENMAIER

WIENER-NEUSTADT OEDENBURG

SEMMERING

NEUSIEDLER
LAKE

RAAB

AGS

Leitha R.

Raab R.

KORMEND VÁSARHELY
STEINAMANGER SARVAR
GUNS
SARVAR

HIDVÉG

PAPA

STYRIAN ALPS

GRAZ

ITALY
ARCHD. JOHN
(75,000)

CARINTHIA

GNAS

HUNGARY

LAKE BALATON

VILLACH HILLERMARKT

KLAGENFURT

TARVIS

VIII A. GIULAY
IX I. GIULAY

(59,000)

MARBURG

Drave R.

Mur R.

CROATIA

Drave R.

UL
UDINE
(18,000)

LAYBACH

PALMANOVA CARNIOLA

RANN

AGRAM

Save R.

ILLYRIA

TRIESTE

ISTRIA

FIUME

DALMATIA

ATIC
A

XX Stoichewich (8,000)
XXX
XI MARMONT (10,000—plus 4,000 troops in garrisons)
confronted each other approximately 100 miles
south-southeast of Fiume.

trian offensive began prior to April 15, the concentration should be further to the rear in the neighborhood of the Donauwörth. The movement toward Donauwörth began as planned. Davout called in the units of his corps, crossed the Danube at Ratisbon, left a garrison in the town, and moved toward Ingolstadt, preparing to continue to Donauwörth. Napoleon had been communicating to Berthier at Strasbourg via manual telegraph and mounted courier. The manual telegraph was the quicker of the two; however, fog caused a delay in sending a message by telegraph on April 10. When it arrived, it was taken out of sequence in the series of messages that Berthier had received by courier. Not knowing of the mistake, Berthier believed that Napoleon wanted Davout and his corps at Ratisbon.[5]

The result was that Berthier ordered Davout to countermarch from Ingolstadt back to Ratisbon. He also ordered Lefebvre, who was falling back before the Austrians, to halt on the Isar River. Davout, who understood Napoleon's concept of operations better than Berthier, protested but to no avail. The result was that instead of the VII and III Corps concentrating to meet the Austrians, they were moving apart and in danger of being destroyed in detail.

Berthier was in a difficult position. It was easy in peacetime to follow orders, but in war, events moved quickly and commanders were expected to show initiative. Berthier was not a commander and had difficulty grasping Napoleon's concept of the campaign. Davout understood. The inefficient command system was soon corrected. Napoleon left Paris on April 13 to take personal command of the Army of Germany, arriving at headquarters at Donauwörth on April 17. The French were very lucky.

The Austrians had crossed the Inn River on April 10 on a broad front between Passau and Salzburg. An Austrian division under Joseph von Dedowich was detached from Rosenberg's IV Corps to besiege the French garrison at Passau. Another division, under Franz von Jellachich (Jellacic on maps), was detached from Hiller's VI Corps to cross at Salzburg and advance toward Munich. What saved the French army was the slow marching speed of the Austrians. The Austrians had not been able to reduce their baggage trains as the French had done. There were too many carriages and wagons owned by Austrian officers clogging the roads. In addition, the Austrians were still reluctant to forage as the French did, and their reliance on supply trains slowed the advance. Consequently, it took the Austrians six days to advance sixty miles from the Inn River to the Isar.

On April 16, there was fighting between Austrian units and Lefebvre's VII Corps as the Austrians approached the Isar. The Bavarians were driven back, and the Austrians secured crossings over the Isar at Landshut and Mossburg. Napoleon realized that the bulk of the Austrian army was on the Isar. His plan of operation was to wrest the initiative from Charles and then destroy him. He could use part of his army to pin the Austrians frontally while another part would turn the Austrian left and throw them into the Danube.

Napoleon planned to unite Davout and Lefebvre and use them as a pinning force while he used the rest of his army (II and IV Corps) as his *masse de manoeuvre* to turn the Austrian left wing (see Appendix A for an explanation of terms). To this end, Napoleon ordered Davout and Lefebvre to join south of Ratisbon in the vicinity of Abensberg. This would form the *masse primaire*. Massena and Oudinot would concentrate at Pfaffenhofen to form the *masse de manoeuvre*. The VIII Corps, the Imperial Guard, and other units would move to Ingolstadt to form a reserve ready to support either *masse*. Napoleon was still unaware of the total size of the Austrian army and its exact order of battle. He had confidence in his forces. He also held his enemy in deep contempt. If Napoleon had been facing a faster moving opponent, the French would have been in serious danger. Neither commander knew the exact location of the opposing forces. The two armies were closing toward each other over an extremely broad front. What followed was a series of confusing engagements in which the rival commanders would gain only an imperfect picture of the situation.

Napoleon has been criticized for developing a scheme of operations without knowing the exact whereabouts of the enemy forces and for making false assumptions about the enemy. However, that is the very nature of modern war. Commanders must make estimates with imperfect intelligence. Napoleon in 1809 acted no differently than he had in 1806. In the latter case, during the Jena campaign, Napoleon believed that Davout was opposed by a small force, when in reality he faced the main Prussian army. What gave Napoleon the edge in past efforts was the archaic nature of his opponent's armies. Napoleon and his commanders were accustomed to operating on dispersed fronts while their opponents were not. In 1809, however, the opposition had changed. Napoleon faced an army of more than 200,000 troops similarly organized to his own army. For the first time, the French would be fighting an opponent who was trying to fight like them, on a dispersed front with army corps.

What happened was a clash by various units along the dispersed front. The different units fought a series of sequential and continuous tactical actions whose effect on the campaign was cumulative. In short, it was distributed maneuver fought on a wide front. This phenomenon predates similar battles fought by Moltke, known as the Battles of the Frontier, in 1866 and 1870. In 1809, as in 1866 and 1870, the army commanders were unaware of the location of the enemy and the nature of affairs and were dependent on the tactical ability of their subordinates.

Archduke Charles hoped to isolate and destroy Davout before rein-

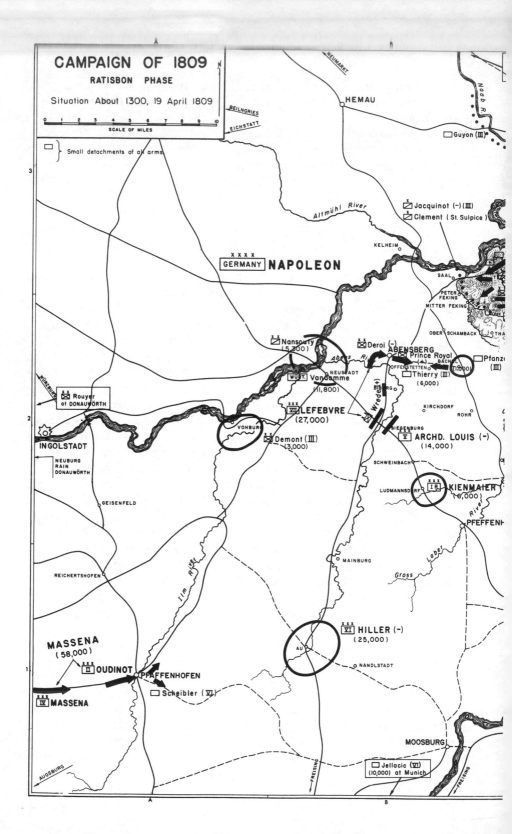

CAMPAIGN OF 1809

RATISBON PHASE

Situation About 1300, 19 April 1809

SCALE OF MILES
0 1 2 3 4 5 6 7 8 9 10

☐ Small detachments of all arms

NEUMARKT

HEMAU

☐ Guyon (III)

Naab R.

Altmühl River

⊠ Jacquinot (-)(III)
⊠ Clement (St. Sulpice)

KELHEIM

XXXX
GERMANY NAPOLEON

SAAL

PETER FEKING
MITTER FEKING

OBER SCHAMBACK

⊠ Nansouty (5,300)

⊠ Deroi (-)   ABENSBERG
Prince Royal

☐ Pfanz (III)

WÜRZBURG

☐ Rouyer at DONAUWÖRTH

WÜRT. Vandamme (11,800)

NEUSTADT

OFFENSTETTEN

BACH

☐ Thierry (III) (6,000)

XX (11,000)

Abens Riv.

Wrede (+)

☆
INGOLSTADT

NEUBURG
RAIN
DONAUWÖRTH

VII LEFEBVRE (27,000)

VOHBURG

☐ Demont (III) (3,000)

KIRCHDORF

ROHR

SIEGENBURG

V ARCHD. LOUIS (-) (14,000)

SCHWEINBACH

GEISENFELD

LUDMANNSDORF

IR KIENMAIER (6,000)

Laber River

PFEFFENH

REICHERTSHOFEN

Ilm River

MAINBURG

Gross

Laber

MASSENA (58,000)

II OUDINOT

PFAFFENHOFEN

☐ Scheibler (VI)

AU

VI HILLER (-) (25,000)

NANDLSTADT

IV MASSENA

MOOSBURG

FREISING

☐ Jellacic (VI) (10,000) at Munich

FREISING

AUGSBURG

A   B

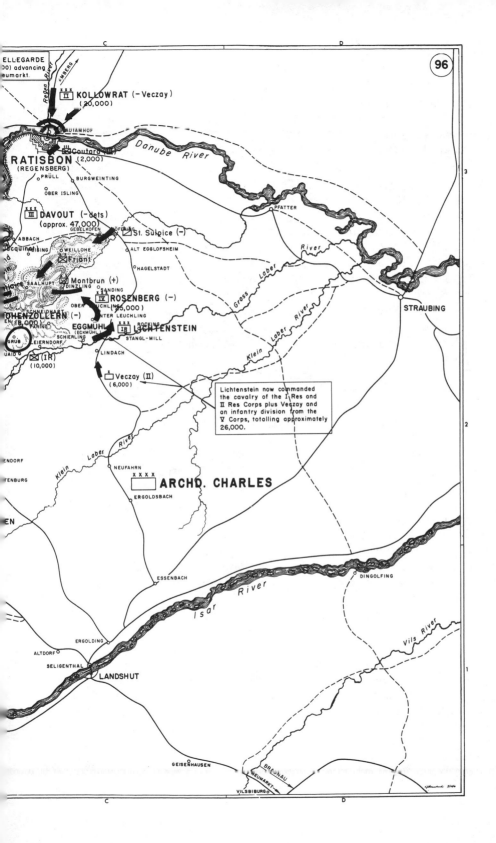

ELLEGARDE
00) advancing
eumarkt.

EMBERG

Regen River

☰ KOLLOWRAT (−Veczay)
(20,000)

ADIAMHOF

Danube River

☷ Coutard (2,000)

RATISBON
(REGENSBERG)

PRÜLL

BURGWEINTING

OBER ISLING

PFATTER

☰ DAVOUT (−dets)
(approx. 47,000)

GEBELKOFEN

DÖFERING

☐ St. Sulpice (−)

ABBACH

ALT EGGLOFSHEIM

acquine AISING

WEILLOHE

HAGELSTADT

River

Gross Laber

Friant

SAALHUPT

☐ Montbrun (+)

DINZLING

SANDING

STRAUBING

HAYE

OBER

☳ ROSENBERG (−)

OHENZOLLERN (−)

LUCHLING

(25,000)

EN (18,000)

SCHEIDMA

UNTER LEUCHLING

PARING

EGGMÜHL

ROCKING

Klein Laber River

GRUB

(ECKMÜHL)

☲ LICHTENSTEIN

UAID

SCHIERLING

EIERNDORF

STANGL-MILL

☒ (IR)
(10,000)

LINDACH

Klein Laber River

☐ Veczay (II)
(6,000)

Lichtenstein now commanded
the cavalry of the I Res and
II Res Corps plus Veczay and
an infantry division from the
V Corps, totalling approximately
26,000.

ENDORF

TENBURG

NEUFAHRN

Klein Laber River

EN

XXXX
☐ ARCHD. CHARLES

ERGOLDSBACH

ESSENBACH

DINGOLFING

Isar River

Vils River

ERGOLDING

ALTDORF

SELIGENTHAL

LANDSHUT

GEISENHAUSEN

BRAUNAU

NEUMARKT

VILSBIBURG

forcements arrived. On April 17, he sent Hiller's VI Corps to Au to cover his left flank while the rest of his army moved north and west to approach Ratisbon and intercept Davout. Kollowrat's II Corps from Bohemia was to approach Ratisbon from the north.

On April 19, Davout, after having left an infantry regiment to hold Ratisbon, was moving south, heading toward Abensberg via Saal. Lefebvre had been withdrawing toward Abensberg, slowly followed by the Austrian V Corps and I Reserve Corps. On the same day, units of Charles' center, consisting of the II Corps and IV Corps, ran into the flank and rear guards of Davout's corps near the village of Thann. The area in this part of the Danube river valley is heavily wooded and hilly. The topography of the various battle areas would impede the use of massed artillery and favored the use of infantry in open skirmisher formation. The French were particularly adept at such light infantry tactics, and the Austrians were not. Therefore, the French would usually have a distinct tactical advantage in this and succeeding engagements. A battle developed around Thann in which the divisions of Generals Saint-Hilaire, Friant, and Montbrun drove the Austrians back.

The road clear, Davout linked up with Lefebvre's corps at Abensberg by the end of the day. Meanwhile, Lefebvre had fought a skirmish outside of Abensberg and kept an Austrian brigade at bay. At the same time, Massena and Oudinot were approaching Pfaffenhofen. Vandamme's VIII Corps was also approaching Abensberg from the east.

On the evening of April 19, the two armies faced each other along a fifty-mile front from Pfaffenhofen to Ratisbon. The situation was as follows: Charles' army from left to right had a detachment near Pfaffenhofen; VI Corps at Au; V Corps and I Reserve Corps south of Abensberg; III Corps, IV Corps, and I Reserve Corps between Thann and Eckmühl (Eggmühl on maps); and II Corps advancing on Ratisbon from the north. Napoleon had Massena and Oudinot advancing on Pfaffenhofen and Lefebvre and Vandamme before Abensberg. Davout's III Corps extended from Abensberg to Abbach, and a regiment of III Corps under Colonel Coutard held Ratisbon.

The units of Napoleon's army were now within effective supporting distance. Napoleon and Charles could fight each other with approximate parity in numbers. Charles had clearly failed to prevent the assembly of the French army and had failed to destroy it in detail. Kollowrat was advancing on Ratisbon, but as long as it was held by the French, Austrian reinforcement from the north bank would be difficult. The initiative was passing to Napoleon.

Napoleon's operational plans matured. He planned to attack the Austrian center while his own right would envelop the Austrian's left via Moosburg and Landshut. Davout's III Corps would form the left anchor of the French line and remain passive. The task of breaching the Austrian line was given to the VII and VIII Corps and a task force formed from the infantry divisions of Morand and Gudin, Jacquinot's light cavalry brigade, and the heavy cavalry divisions of Nansouty and Saint-Sulpice. This force, totaling 25,000 men, was placed under the command of Marshal Jean Lannes, who had just arrived from Spain. Lannes was a superb offensive commander, matching Ney in personal bravery, and made the perfect advance guard commander.

Meanwhile, Massena, who was given operational command of the *masse de manoeuvre*, would turn the Austrian left and head for the Austrian crossings over the Isar to cut them off. Napoleon thought this would trap and destroy most of Charles' army, since the Emperor mistakenly believed most of Charles' army was farther south and not facing Davout.

Unaware of the approaching storm, Charles was shifting his strength to his north, while Napoleon was shifting to his forces to the south. The object for the Austrians was to take Ratisbon and defeat Davout. The actions on the following day, April 20, are collectively referred to as the Battle of Abensberg, although it was not a single battle but, rather, a series of tactical engagements along a broad front twenty-five miles wide.

Archduke Louis and his V Corps were the intended targets of the French attack. Louis had 14,000 troops overextended between Siegenburg and Kirchdorf. To his right was a reinforced brigade from III Corps at Bachel. Echeloned behind him extending to Pfeffenhausen were Kienmayer and Hiller. Against Louis were 25,000 men of Lannes' task force and 30,000 of the VII and VIII Corps. Hiller and Kienmayer were within supporting distance, but there was little cooperation among these corps commanders and no one showed initiative, something needed to fight a decentralized battle along an extended front. Archduke Charles seemed preoccupied with his right and was too far away to effectively control the corps on his left.

Lannes struck like a thunderbolt at 9:00 A.M. and shattered the detachments of the III Corps to Louis' right. Lefebvre's Bavarians and Vandamme's Württembergers attacked as well, smashing Louis' corps. Streaming back in flight, they carried part of Kienmayer's II Reserve and Hiller's VI Corps with them. Hiller took effective control of the wing and ordered what was left of V Corps as well as his own VI and Kienmayer's II Reserve Corps to retreat back toward Landshut. The French succeeded in sundering the Austrian left wing from the center. The Austrian losses totaled 6,700.[6]

Colonel Coutard, commanding the French garrison at Ratisbon, was surrounded and besieged by Kollowrat's II Corps from the north and Lichtenstein's I Reserve Corps from the south. The French were unable to destroy the bridge at Ratisbon. Hopelessly outnumbered, Coutard surren-

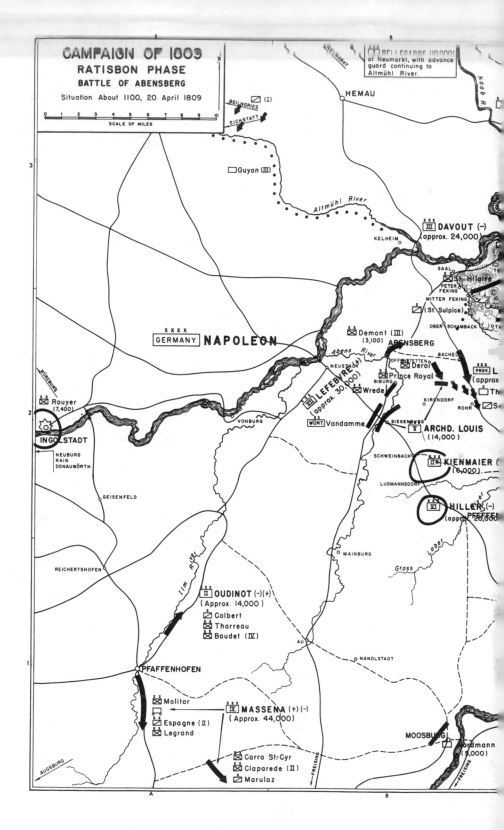

CAMPAIGN OF 1809
RATISBON PHASE
BATTLE OF ABENSBERG
Situation About 1100, 20 April 1809

SCALE OF MILES
0 1 2 3 4 5 6 7 8 9 10

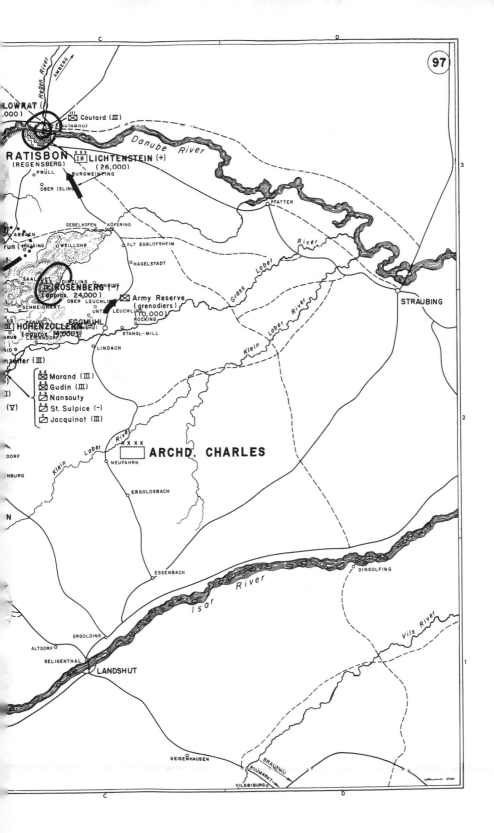

LOWRAT (
,000)

⊠ Coutard (Ⅲ)

STADTAMHOF

Regen River

AMBERG

C

D

Danube River

RATISBON
(REGENSBERG)

ⅠR LICHTENSTEIN (+)
(26,000)

PRÜLL

BURGWEINTING

OBER ISLING

PFATTER

ABBACH

GEBELKOFEN

KÖFERING

run (HESSING

OWEILLOHE

ALT EGGLOFSHEIM

Grass Laber River

HAGELSTADT

SAALHUET

DINZLING

SCHNEIDHART

ROSENBERG ( )
(approx. 24,000)

OBER LEUCHLI

⊠ Army Reserve
(grenadiers)
(10,000)

STRAUBING

UNTE LEUCHLI

ROCKING

HOHENZOLLERN

EGGMUHL

STANGL-MILL

Klein Laber River

LEIENDORF

GRUB

LINDACH

nzeller (Ⅲ)

⊠ Morand (Ⅲ)
⊠ Gudin (Ⅲ)
⊠ Nansouty
⊠ St. Sulpice (-)
◿ Jacquinot (Ⅲ)

(Ⅴ)

Laber River

XXX ARCHD. CHARLES

DORF

NEUFAHRN

NBURG

Klein Laber River

ERGOLDSBACH

N

ESSENBACH

DINGOLFING

Isar River

Vils River

ERGOLDING

ALTDORF

SELIGENTHAL

LANDSHUT

GEISENHAUSEN

BRAUNAU

NEUMARKT

VILSBIBURG

C

D

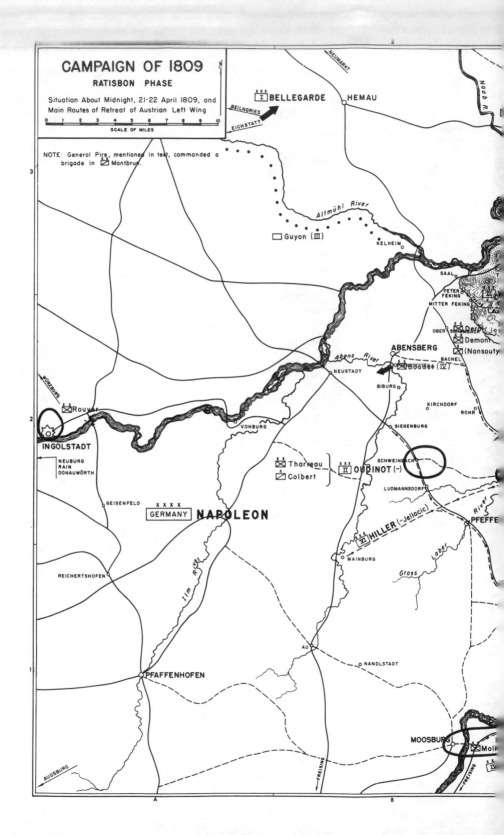

CAMPAIGN OF 1809
RATISBON PHASE

Situation About Midnight, 21-22 April 1809, and
Main Routes of Retreat of Austrian Left Wing

0 1 2 3 4 5 6 7 8 9 10
SCALE OF MILES

NOTE: General Pire, mentioned in text, commanded a
brigade in ☒ Montbrun.

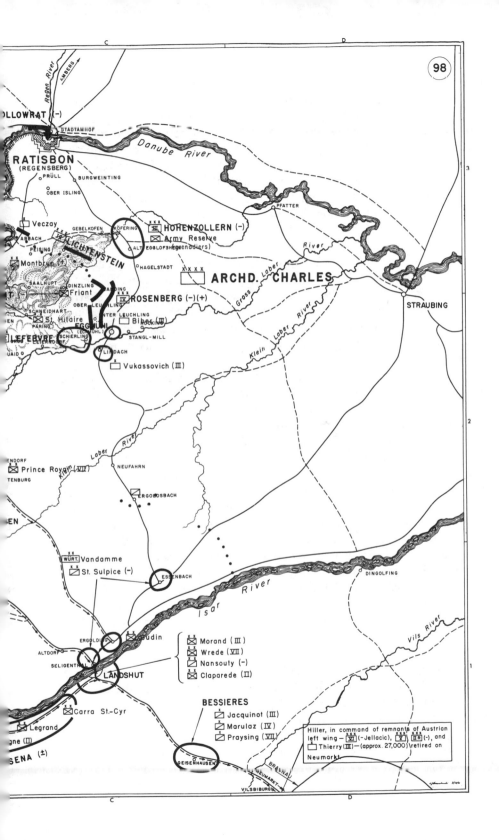

Regen River

LIMBERG

OLLOWRAT (–)

STADTAMHOF

Danube River

RATISBON
(REGENSBERG)

PRÜLL

BURGWEINTING

OBER ISLING

PFATTER

Veczay

GEBELKOFEN

KÖFERING

HOHENZOLLERN (–)
Army Reserve
(ALT EGGLOFSHEIM Grenadiers)

ARBACH

PEISING

Montbrun (+)

HAGELSTADT

River

STRAUBING

SAALHUPT

DINZLING

Friant

Gross Laber

ARCHD. CHARLES

SCHNEIDHART

OBER LEUCHLING

Laber River

St. Hilaire

PARING

UNTER LEUCHLING

ROSENBERG (–)(+)

LEFEBVRE

EGGMÜHL

SCHIERLING

BIROKIN

Biber (III)

STANGL-MILL

Klein

UAID

LINDACH

Vukassovich (III)

Laber River

ENDORF

Prince Royal (VII)

NEUFAHRN

Laber River

TENBURG

SEN

Klein

ERGOLDSBACH

WÜRT Vandamme

St. Sulpice (–)

ESSENBACH

Isar River

DINGOLFING

Vils River

ERGOLDING

Gudin

ALTDORF

SELIGENTHAL

Morand (III)
Wrede (VII)
Nansouty (–)
Claparede (II)

LANDSHUT

Carra St.-Cyr

BESSIERES

Jacquinot (III)
Marulaz (IV)
Praysing (VII)

Legrand

gne (II)

SENA (±)

GEISENHAUSEN

BRASNAI

NEUMARKT

VILSBIBURG

Hiller, in command of remnants of Austrian
left wing — VI (–Jellacic), V, IIR (–), and
Thierry (III) — (approx. 27,000) retired on
Neumarkt

dered. Charles, having lost contact with his left wing, now gained contact with his right wing and won control of a vital crossing over the Danube.

For the next day, April 21, Lannes, Massena, Vandamme, and also Marshal Jean-Baptiste Bessières, commanding another composite force, pursued the defeated Austrian left wing toward Landshut. Under the effective command of Hiller, the left wing managed to cross the Isar at Landshut before the French could cut them off. Meanwhile, Lefebvre moved to support Davout, who was still facing the Austrian center, which was being reinforced by Kollowrat. Charles had pulled back his line, with his right before Abbach on the Danube and his left at Eckmühl on the Gross Laber River. The front was ten miles long.

Charles had been shifting his forces to his right at the time Napoleon was moving against Louis. Consequently, a false sense of reality existed in Napoleon's mind. He believed he had defeated the bulk of Charles' army rather than a single wing. Napoleon sent off Bessières to pursue Hiller, believing he had beaten Charles' army. Napoleon was also unaware that the bridge at Ratisbon was intact and that Kollowrat was crossing the Danube.

By sending half his army toward Landshut because he thought he had defeated most of Charles' army, Napoleon left Davout and Lefebvre increasingly isolated facing the central Austrian column which was being reinforced from the right. Davout and Lefebvre collectively had 36,000 troops while Charles had 75,000.

Charles resolved to attack Davout and Lefebvre on April 21. The Austrian main effort would be on their right against Davout's weak left, with the intention of driving the French from the south bank of the Danube and away from their communications that ran Abensberg-Ingolstadt. The attack would be made from the right by Kollowrat's II Corps and Lichtenstein's I Reserve Corps. Hohenzollern's III Corps and Rosenberg's IV would hold the Austrian left. Meanwhile, Bellegarde's I Corps was approaching Ratisbon from the north. Davout reported the Austrian buildup on his front to Napoleon, telling him, "I will hold my position—I hope."[7] It was not until 2:00 A.M. on July 22 that Napoleon learned Davout was facing the bulk of Charles' army. Napoleon then decided to move northeast to attack the left flank of Charles' forces fighting between Abbach and Eckmühl.

The Battle of Eckmühl was fought April 21–22. Davout and Lefebvre fought the Austrians to a standstill on the first day and into the second. The moment for an Austrian victory had passed. By late afternoon on April 22, the lead elements of Lannes' corps approached Eckmühl from the south, and the rest of Lannes' and Massena's corps extended to the south in a column twenty-four miles long. Oudinot's II Corps was approaching from the west. This would bring 100,000 French troops against 75,000 Austrians.

Davout and Lefebvre attacked the Austrians frontally while Napoleon's envelopment began. Lannes' corps began an attack against Eckmühl. The Austrian left, consisting of Rosenberg's IV Corps, was driven from Eckmühl, and as the attack of Lannes' corps developed, Charles realized he was in danger of being rolled up from the left and thrown into the Danube. He ordered a retreat. By evening Napoleon, concerned about the confusion of a night pursuit, ordered a halt. Almost all of the French forces were exhausted from six days of constant marching and fighting.

During April 23, Charles was able to cross his army to the north bank of the Danube over the Ratisbon bridge, leaving a 6,000-man garrison in the town. Hoping to pursue the retreating army, the troops of Lannes' corps attacked Ratisbon many times. An assault led in person by Marshal Lannes finally carried the fortifications. The French took the town, but the Austrians still controlled the bridge. A French garrison would have to be kept at Ratisbon to prevent an Austrian threat to the south bank. In all, the battles from Thann through Ratisbon cost the Austrians 30,000 casualties;[8] the French lost about half that number.

The Austrians had suffered a major strategic defeat, and their hopes of winning an offensive campaign were shattered. Save for a revolt in the Tyrol, there was no effective uprising against the French. Their hope of catching and destroying the separate formations of the French army had failed. What was worse, Napoleon had wrested the strategic initiative from the Austrians. Two-thirds of Charles' army had been driven north of the Danube, and the remaining third was retreating along the southern bank. Napoleon's main army was concentrated, and the way to Vienna appeared open. Napoleon could advance south along the Danube to Vienna or try to cross north of the river to pursue Charles.

Charles' army was outmaneuvered and beaten in the series of engagements from April 19 to 24, and that series pointed to the way war would be conducted in the future. Two comparatively large armies, 165,000 against 209,000, both organized into corps, produced a remarkably symmetrical operation. Both sides maneuvered according to a preconceived plan of operations, and the rival forces clashed on a broad front averaging seventy miles in length. The size of the armies and the corps structure enabled these broad deployments, creating sequential and continuous engagements from April 19 to 24. The engagements were totally different from the battles of Austerlitz or Friedland, battles fought on narrow fronts and lasting only one day. Napoleon could personally view those battles. In the engagements in 1809, there was no single battle, and neither commander physically could view the entire operational front.

This type of warfare breeds confusion. Accurate reconnaissance is diffi-

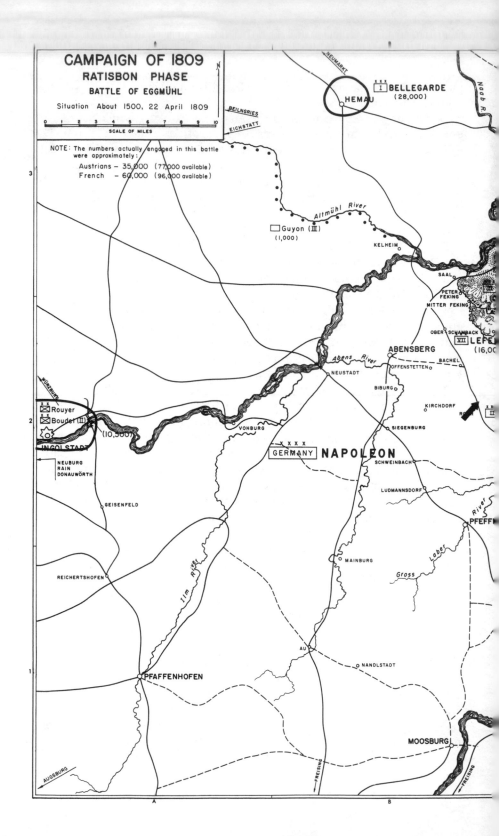

CAMPAIGN OF 1809
RATISBON PHASE
BATTLE OF EGGMÜHL

Situation About 1500, 22 April 1809

SCALE OF MILES
0 1 2 3 4 5 6 7 8 9 10

NOTE: The numbers actually engaged in this battle
were approximately:
Austrians — 35,000 (77,000 available)
French — 60,000 (96,000 available)

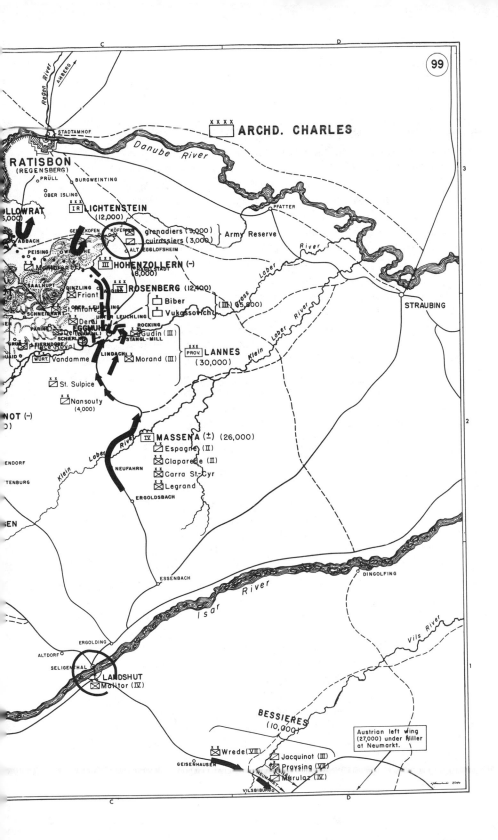

ARCHD. CHARLES

Regen River

AMBERG

STADTAMHOF

Danube River

RATISBON
(REGENSBERG)

PFATTER

PRÜLL   BURGWEINTING

OBER ISLING

KOLLOWRAT
(3,000)

IR LICHTENSTEIN
(12,000)

GERZKOFEN

ABBACH

KÖFERING   grenadiers (9,000)

Army Reserve

PEISING

ALT EGGLOFSHEIM   cuirassiers (3,000)

River

III HOHENZOLLERN (-)
(8,000)

NAGELSTADT

Laber   River

MOKING R.

SAALHUPT

DINZLING

ROSENBERG (12,400)

STRAUBING

Friant

OBER FUCHING

SCHNEIDHARD

St. Hilaire

Biber

(III) 65,000)

St.

ST. Hilaire

UNTER LEUCHLING

Vukassovich

Gross

Laber   River

FARING

Deral

(Deiflhut)   ROCKING

Gudin (III)

SCHIERL

STANGL-MILL

FIERNDORF

Klein   Laber   River

WURT. Vandamme

LINDACH

Morand (III)

PROV. LANNES
(30,000)

St. Sulpice

Nansouty
(4,000)

NOT (-)
)

Laber   River

ENDORF

IV MASSÉNA (±) (26,000)

Espagne (II)

NEUFAHRN

Claparède (II)

EN

TENBURG

Carra St.-Cyr

Legrand

ERGOLDSBACH

Klein   Laber   River

ESSENBACH

DINGOLFING

Isar   River

Vils   River

ERGOLDING

ALTDORF

SELIGENTHAL   LANDSHUT

Molitor (IV)

BESSIÈRES
(10,000)

Austrian left wing
(27,000) under Hiller
at Neumarkt.

Wrede (VII)

GEISENHAUSEN

Jacquinot (III)

Preysing (VII)

NEUMARKT

Marulaz (IV)

VILSBIBURG

cult if not impossible, and army commanders gain only an imperfect picture of the actual situation and must depend upon the initiative and abilities of their subordinate commanders. It is understandable that Napoleon thought he had chased the bulk of Charles' army over the Isar at Landshut, just as he thought he faced the main Prussian army at Jena in 1806. It can be argued that continuous and sequential operations on a similar broad front were apparent in the double battles of Jena-Auerstadt. The difference, however, is that the Prussian army of 1806 lacked the organizational structure to perform distributed maneuver. The movements of that army immediately prior to the battles of Jena-Auerstadt were accidental rather than by any operational design. The Prussians were already in retreat and were trying to avoid the French, not fight a battle. The antiquated Prussian army could not have provided the symmetrical situation that existed in 1809, and this was why, in part, the Prussians collapsed, and the Austrians in 1809 did not.

Napoleon was comfortable with this type of situation and was confident in the abilities of his subordinate commanders. Archduke Charles, on the other hand, was not used to this at all, and his nerve broke. The loss of the Prussian alliance and the redeployment of the Austrian army in March seemed to have broken his morale, and his pessimism and lack of effective command sapped energy from the Austrian offensive. Charles never got a clear picture of the battlefield and never was able to press his advantages. His lack of effective command was compounded by the inability of his corps commanders to aggressively handle their corps. Charles' pessimism colored his appraisal of the situation, and he reported that his army was disorganized and demoralized and that Austria should sue for peace.

Considering that corps had just been created in the Austrian army, corps commanders and staffs were relatively inexperienced. Charles may have been asking too much to have expected a better performance; it would take time to close the gap between the French and Austrian armies. But all was not black. There was in reality, some grounds for Austrian optimism. Although defeated, the Austrian army was not destroyed, and the casualty total was but 30,000 out of a total strength of 209,000. The Austrian right wing (I and II Corps) had barely been touched and several of the other corps had been only moderately engaged. In marked contrast to the battles in 1805–1807, there was no army-wide collapse or rout. The new corps structure ensured that the Austrian army would recover and fight again; these battles were only a phase in a longer war.

In many ways, the Austrian army was like the Army of the Potomac in the American Civil War. Both armies often fought under mediocre commanders and lacked brilliance in maneuver, but they had a dogged resilience. Similarly, the battles of Thann-Ratisbon were like battles fought during the 1862 Peninsular campaign in the American Civil War, a campaign characterized by a series of engagements in which the Confederates wrested the initiative from the Federals and forced them to retreat. The Army of the Potomac in that campaign was dispirited but not broken, and the same can be said of the Austrian army.

Large armies symmetrically organized in corps, fighting indecisive, continuous, and sequential engagements on broad fronts increasingly became the norm as warfare progressed in the nineteenth century and represented a break from the way warfare was previously conducted. This campaign along the Danube represents an example of that progression. Neither Napoleon, Charles, nor perhaps anyone else at the time realized that the nature of war was changing—but it was.

Napoleon now had options as to what to do next. Many have said that he should have launched an all-out pursuit north of the Danube. However, to bring his army over that river would move his center of gravity further away from the Italian theater of operations, and Napoleon was receiving incomplete and disquieting reports of a disaster sustained by the Army of Italy. In addition, he was unclear about the status of his strategic right. Would Archduke John's army suddenly appear on his right flank?[9]

A move north of the Danube would pull him away from Italy and open his line of communication to interdiction from the south. Moreover, Napoleon favored a strategic division of his enemies in Germany and Italy. If Napoleon moved north of the Danube, then the Austrian armies in Germany and Italy would have a better chance to unite. If he advanced south of the Danube to Vienna, he could keep the two enemy armies divided. It is quite evident that Napoleon, as supreme commander, saw the interrelationship of the German and Italian theaters of operation. What he had to know was exactly what was happening in Italy.

CHAPTER SIX

Crisis in Italy

In Italy, the French were surprised by the Austrian offensive. The Army of Italy had been deliberately dispersed to avoid provoking the Austrians and to ease the logistical burden that would result from a premature concentration in eastern Venetia. Eugene de Beauharnais had been lulled into a false sense of security primarily by Napoleon's inability to believe that danger was imminent. Right up to the end of March, Napoleon was convinced that hostilities would not begin until the middle of May.[1] There would still be plenty of time to concentrate the army. Napoleon also believed that Russia would actively support him. Consequently, he believed that any Austrian offensive would be suicidal, especially with Russia threatening Austria's eastern frontiers. Writing to Eugene, he said: "As for me I will remain stationary for all of April, and I do not think the Austrians will attack, especially after the Russians march on Hungary and Galicia."[2] As we have learned, Napoleon was mistaken by placing his trust in the Russian alliance. Eugene was strongly influenced by Napoleon's views, and that shaped both his interpretation of enemy movements prior to the invasion and subsequent actions. There was no urgency in building the Adige defenses; the target date for completion was the end of April.[3] And, as mentioned previously, Napoleon and Eugene delayed in the creation of corps in the army and in the appointment of corps commanders.

Intelligence reports coming into Viceregal headquarters gave a picture of greater danger. One report that Eugene received on March 13 predicted an Austrian invasion of Italy some time between March 25 and April 1.[4] On March 20, it was reported that fourteen Austrian infantry regiments, six cavalry regiments, and twelve *Landwehr* battalions were massing in Carinthia.[5] On March 23, Eugene learned that there were Austrian troop concentrations at Laibach, Klagenfurt, Villach, and Salzburg.[6] The Austrians were also building fortifications at Malborghetto, Mount Predel, and Prewald opposite the Italian frontier.[7] Eugene's agents informed him that "all of Austria is dreaming and talking of war,"[8] and that Austrian agents were spreading rumors within Hapsburg territories that the French armies in Spain had been destroyed and that the Russians were marching to support Austria.[9] However, Eugene clung to Napoleon's views. Writing to Marmont, for example, on March 23, Eugene related that "the Russians are marching against Austria. . . . She will be attacked from all sides. Their plan of campaign is upset even before war is declared."[10]

April 1 came and went without the anticipated Austrian attack. This may have enhanced Eugene's confidence in his stepfather's analysis and in the belief that no offensive would come until May 1. However, the reports sent by Eugene to Napoleon, as well as from other sources, caused Napoleon to reassess the situation. On March 30, he came to believe that an Austrian offensive was likely on or after April 15.[11] Although the Army of Germany was effectively positioned to begin operations, no orders were sent to do so for the Army of Italy. Failure to alert Eugene in time was a grave oversight on Napoleon's part.

Eugene was still operating under the belief that hostilities would not begin until the end of April. Reports of a pro-Austrian riot that had broken out in Tuscany on April 2 did not cause Eugene to reassess his position. Nor did he believe a message from his agent in Trieste, a M. Segurier,[12] who reported that a great number of Austrian troops and artillery were moving out of the city and that a war chest of 4 million gold pieces had arrived. According to Segurier, "hostilities are expected soon."[13] Unfortunately for the French, this report was correct.

Austrian forces were on the move. Leaving two brigades behind, Ignatius Giulay moved his IX Corps of two divisions north to rendezvous with the VIII Corps on the upper Isonzo. By April 9, the Army of Inner-Austria was approaching the frontiers. Chasteler, with a divisional-sized force detached from VIII Corps, moved toward the Tyrol. Early on April 10, as in Germany, the Austrians sent messages to the French outposts along the frontier, informing them that the Austrians were going to advance and "treat as enemies"[14] all who opposed them. There was no formal declaration of war.

The Austrians promptly attacked along a 100-mile front. On the right, Chasteler moved in the Tyrol, heading for Linz and triggering a prearranged revolt. The Bavarian garrisons in the Tyrol were either wiped out by Tyrolean partisans or driven away. A French column of 2,000 troops moving through the area at the time was cut off and forced to surrender at Innsbruck. A second column of French troops was driven south toward Trent.

The main body of Archduke John's army, with the VIII and IX Corps, moved down the Isonzo toward Caporetto. From there the column would advance to Cividale and Udine. To the right of the main column, a regiment attacked the French outpost at Ponteba. While Count Gavassini's brigade, which had been left in Carniola, crossed the Isonzo from Goritzia to cover the main body's left, Stoichewich attacked southward into Dalmatia. The French retreated before him.

Eugene was at Udine when the first reports of the offensive arrived.

What should he do? There were the two operational plans developed by Napoleon, one offensive and the other defensive. Offensive action was out of the question at the moment. The defensive plan called for a withdrawal behind the Adige line. However, it is now axiomatic that no plan survives the first contact with an enemy.[15] Not all contingencies can be expected or planned. Information is imperfect, and the whereabouts and intentions of the enemy, as well as the location of one's own troops at any given time, are uncertain. To perform in this environment, a commander must act on uncertainty, be willing to take risks, and not become overwhelmed by the pressures of war. In the final analysis, success in war depends on character, whether one can take the strain or not. This is something that no one can teach or train even in modern staff colleges. Knowledge of any system of war or doctrine is useless without intelligence, moral courage, and nerve. Did Eugene de Beauharnais have these qualities?

His last combat experience was as a captain at Marengo. He had showed courage and pluck, but his past responsibilities were limited to that of commanding a cavalry troop. Now he was an army commander, responsible for a theater of operations consisting of hundreds of square miles, including thousands of troops and millions of civilians. Could he handle the perils of command? Was knowledge of Napoleon's methods of war enough?

Initially, Eugene ordered the defensive plan into operation. Garrisons were to hold Osoppo and Palmanova, and the rest of the army was to fight a delaying action back to the Adige. However, this operation was almost immediately abandoned. Napoleon's plan assumed a major enemy offensive coming from east to west toward the Adige. What was not anticipated was the uprising in the Tyrol. As news of the massacre of the Bavarians and the rout of French troops spread from the Tyrol, so too did the reported strength of Chasteler's division and the Tyrolean partisans.[16] The picture that developed in Eugene's mind was that of a two-pronged invasion with John attacking from the east and Chasteler sweeping down from the Alps behind him. In such a case, it seemed that Napoleon's "notes" had run out.

The young Viceroy was pressured not to withdraw to the Adige. Military planners can blithely concede home territory if it means gaining a later military advantage (as Alfred von Schlieffen did in the twentieth century). Napoleon did the same by advocating a withdrawal to the Adige. Eugene discovered, however (as did the executors of Schlieffen's plan, Helmuth von Moltke the younger and General Max von Prittwitz), that in reality the abandonment of home territory to the enemy was too bitter to contemplate. Both official and private citizens begged the Viceroy not to abandon them to the mercy of the enemy. Eugene's own generals grumbled that they were not accustomed to retreating before the Austrians. As commanders had done before him and would do after him, Eugene changed his plans to meet what he perceived to be the requirements of the moment. Unlike Berthier in Germany, he would not adhere slavishly to a plan that he considered overtaken by events.

Believing himself to be outnumbered and positioned between two converging forces, Eugene's solution followed the spirit of Napoleonic principles but not the letter of the defensive "note." Eugene would rely on interior lines to crush one opponent, then the next. The Viceroy planned to fall back only until he had gathered enough forces to turn and crush John's army; then he would turn north to meet Chasteler.[17] The theory was easy; the execution would be difficult.

For the plan to work, the Tyrolean flank had to be secure to allow Eugene to concentrate against John. The Viceroy ordered General Baraguey d'Hilliers to leave Venice and go to Trent to take over the defenses of the upper Adige. Eugene also ordered the 2d Italian Infantry Division, the French 112th Infantry Regiment, and a French dragoon regiment there to give Baraguey d'Hilliers something to work with. Eugene sent an order to Marmont to attack north from Dalmatia to take off some of the pressure from the Italian front, but this was a forlorn hope since Marmont was at that time withdrawing to Zara.

Eugene had ordered the divisions that were east of the Tagliamento to fall back while he ordered the others, save for those going to Trent, to the east. Eugene's divisions that had originally been on the Isonzo fell back, joining new forces as they retreated. On April 14, Eugene was behind the Livenza River at Sacile with five infantry divisions and a light cavalry division. Another French infantry division, Lamarque's, was at Vicenza sixty miles away, and Pully's division of dragoons was seventy miles away at Padua.

Eugene calculated that the divisions of Lamarque and Pully could be at Sacile by April 16. John's army was approaching the Tagliamento and would cross it the next day. Eugene figured that the Austrians would approach the Livenza at Sacile late on April 15. Consequently, the Viceroy decided that with the anticipated arrival of Lamarque and Pully he would have eight divisions, strong enough to offer battle at Sacile. So confident was Eugene of the arrival of extra divisions that he wrote to Napoleon on April 14 reporting that Lamarque's division *had already arrived* and that Pully's would arrive the next day.[18]

On April 14, Eugene deployed his army five miles north and south of Sacile along the east bank of the Livenza. Seven miles due east of Sacile was the village of Pordenone, where Eugene sent an advance guard consisting of a light cavalry regiment and an infantry regiment under General Sahuc to look out for the approach of John's army.

It rained heavily on the night of April 14–15. The rain muddied the roads severely, delaying the march of Lamarque and Pully. It also made the French scouting patrols lethargic; besides, they were only fighting the Austrians. John's army moved up the road from Valvasone toward Pordenone. Marching in succession came the Advance Guard, VIII Corps, and IX Corps. With forces flanking this column, John had close to 44,000 troops. The Advance Guard consisted of a mixed division of infantry, cavalry, and artillery under Feldmarschall-Leutnant Johann Freiherr Frimont von Palota, an energetic officer.

Pressing his men hard, Frimont's cavalry located the French at Pordenone early on the morning of April 15. Frimont organized his forces for a surprise assault, and they were undetected by their enemies. In a neatly executed tactical maneuver, the Austrians routed the French cavalry and surrounded the French 35th Infantry Regiment, forcing its surrender. The peacetime reforms of the Austrian army were taking hold.

Pordenone was seven miles due east of Sacile and was connected to that town by the main east-west road. Midway between Pordenone and Sacile was the village of Fontana-Fredda. Running northeast from Fontana-Fredda were the villages of Ronche and Rovoredo. Extending northwest from Fontana-Fredda were the villages of Vigonovo and Ranxan. South of the main road were the villages of Talponedo, Porcia, and Palse. The terrain north of the main road was open while that south of the main road was broken, cut up by a series of streams and ravines. Three miles south of Sacile and on both banks of the Livenza was the town of Brugnera.

After the victory at Pordenone, Frimont's Advance Guard occupied the villages of Talponedo, Palse, and Porcia, effectively securing the approaches to Pordenone south of the main road. A mixed brigade under Colonel Volkmann occupied Rovoredo on the right while another smaller brigade covered the far Austrian left at La Motta farther to the south. John kept the rest of his army massed around Pordenone. Fontana-Fredda was garrisoned by a French infantry regiment and a light cavalry brigade, while the rest of the French army was behind the Livenza. Both armies were within striking distance of each other. The rest of the day was spent in reconnaissance and preparation. John's army, without the flanking forces, totaled 40,000 men, including 3,000 cavalry. Eugene had 36,000, which included just under 2,000 cavalry. John could expect no reinforcements whereas Eugene could. It would be prudent for the Austrian commander to be cautious; trying to force a river crossing against a foe who nearly matched his own strength could be very dangerous. Eugene had every reason to await further reinforcement, yet he threw caution to the winds and decided to attack with the forces he had. Why?

Eugene was under the impression that time was critical. The reports from the Tyrol falsely indicated that Chasteler had 20,000 troops and was

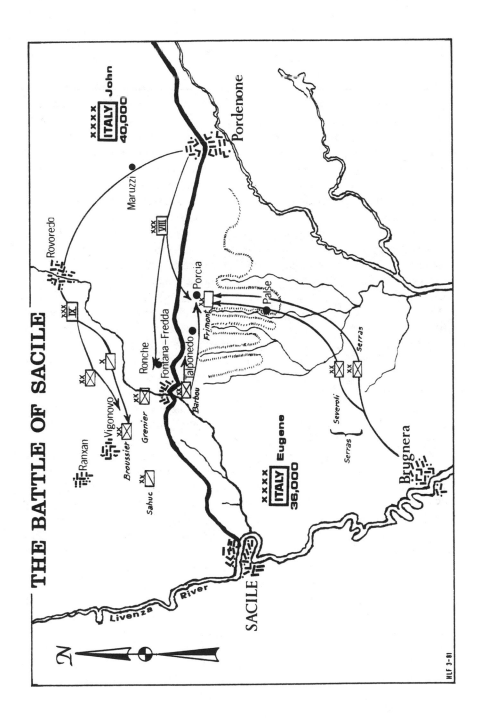

THE BATTLE OF SACILE

on the verge of descending the Alps,[19] so Eugene believed that he would have to quickly destroy John to meet the threat from the north. It seems incredible in hindsight that he thought he could so rapidly defeat John with the forces at hand. What seems evident is that Eugene, along with Napoleon and the rest of the French high command, had contempt for their opponents, believing that these were the Austrians of 1805. The Army of Italy was on the verge of learning that this was no longer the case; Napoleon would learn this a bit later at Aspern-Essling. Eugene was also under the mistaken impression that Lamarque and Pully would arrive sometime on the afternoon of April 16, and so it was all right to launch an offensive against the Austrians with the forces at hand.

Eugene planned an attack with the major effort south of the main road. Two infantry divisions with some cavalry support were to attack from Brugnera and take the villages of Talponedo, Palse, and Porcia, currently held by Frimont's division. The rest of Eugene's army was to attack from Sacile. Two divisions were to support the right wing by attacking along the main road. One infantry division would cover the French left, and Sahuc's reduced cavalry division would form the army reserve. Once the villages were taken and the reinforcements arrived, Eugene would develop the attack toward Pordenone. Evidently, Eugene expected the Austrians to remain passive, since enemy capabilities were not taken into account.

The French attack began at dawn. By 9:00 A.M., the French right wing, consisting of the divisions of Generals Serras and Severoli, had carried Palse. The French moved on to take Talponedo and Porcia, but here Frimont put up a spirited defense and counter-attacking, threw back the French attack wing. Reorganizing, the French returned to the attack, reinforced by General Barbou's infantry division, which had advanced from Sacile. John, in the meantime, fed in reinforcements. Albert Giulay with two brigades of his VIII Corps reinforced Frimont to continue the fight around the villages. Fighting around Porcia became particularly heavy, the village being taken and retaken several times. Eventually the French took the villages, but at a heavy cost to both sides. Although pushed back, Frimont and Giulay continued to put up a bold front.

As the French right wing advanced, the center and left were compelled to keep pace. Grenier's division moved through Fontana-Fredda to Ronche. The French left wing, consisting of Broussier's division, was forced to move out of the villages of Ranxan and Vigonovo and into the open plain between Vigonovo and Fontana-Fredda. As they did so, the entire French army emerged into full Austrian view. John realized that the French were outnumbered and that their left flank was exposed to a turning movement. Frimont and Albert Giulay were to return to the attack to engage the French right and center. Albert Giulay was reinforced by General Gagoli's infantry brigade and Volkmann's mixed brigade. The entire

IX Corps, consisting of an infantry division of two brigades and a cavalry brigade, was to attack and turn Broussier's division. An infantry brigade of grenadiers under Kleinmeyer was to remain in army reserve.

The Austrian counterattack began at 3:30 P.M. The arrival of fresh troops gave the Austrians new heart, and they were also aided by an overall superiority in artillery and cavalry. The French were pushed out of Porcia. Thrown on the defensive, Eugene was forced to commit his last reserve, Sahuc's weakened cavalry division, to support Broussier, but this was not enough to halt the Austrian advance.

Ignatius Giulay's IX Corps advanced to the attack via Rovoredo and began to outflank Broussier. The hoped-for reinforcements of Lamarque and Pully were nowhere in sight, and Eugene realized that with no more reserves left, the battle was lost. He must retreat before his army was overwhelmed.

The divisions of Serras, Severoli, and Barbou retreated to Brugnera, and the divisions of Grenier, Broussier, and Sahuc withdrew to Sacile. Frimont's division and the VIII Corps were exhausted after the fighting and were incapable of pursuit. However, the IX Corps, being fresh, was able to continue to press the French left. At 7:00 P.M., the Austrian cavalry charged Broussier's division. The French infantry, formed into squares, managed to fend off this attack. With night falling, pressure from the Austrians lessened, and the French were able to retreat behind the Livenza. The Austrian army was incapable of further pursuit,[20] and the Battle of Sacile was over. French losses totaled 6,500 and fifteen guns, the equivalent of an entire division. The Austrian casualties came to about 4,000.[21]

Archduke John had won the first major battle for Austria in this war and the first Austrian battlefield success since 1800. Eugene's generalship was very poor, and the Viceroy must take a significant amount of blame for the defeat. However, the battle at Sacile was also a victory for the reforms of the Austrian army. The creation of army corps ensured that the various Austrian formations could effectively coordinate the three combat arms, as evidenced by the combat at Pordenone and by the performance of Frimont's division and of the VIII and IX Corps at Sacile. Eugene himself noted in a report after the battle how effectively the Austrians had used their artillery.[22] Moreover, the reforms enabled John to control the three maneuver units of the Advance Guard, VIII Corps, and IX Corps. The Austrians fought stoutly at Sacile, showing an aggressiveness not seen in their performance in 1805. The course of this battle and the casualties sustained indicated that tactically this was a battle won by attrition. The reforms enabled the Austrians to fight as well as the French. Eugene, with a smaller army, had expended his resources while John still had reserves to commit.

Sacile, however, was not decisive. Although in this battle the Austrians had a numerical superiority, they were still outnumbered in the theater of operations. Had Eugene chosen to remain on the defensive behind the Livenza, John probably would have gone no further. A river crossing against an unbeaten French army would have been hazardous, especially once the French reinforcements arrived to tip the balance. Now that the French had been defeated, there would be no defensive stand on the Livenza, and the Austrians could advance further into Italy.

Disorganization spread in the Army of Italy. The roads had become extremely muddy because of the spring rains and so hampered movement as the retreating columns became intermixed with the army's vehicles. Confusion and nightfall threatened the army with disintegration. Eugene himself had been severely shocked by the defeat; he initially panicked and briefly believed that his army might be routed. However, help was at hand.

The divisions of Lamarque and Pully had been delayed by the spring rains and muddy roads. By the evening of April 16, they were at Conegliano, ten miles west of Sacile. They were directed to cover the retreat, and the proximity of these fresh troops put some heart into the Army of Italy and its commander.

By the evening of April 17, the entire army was behind the Piave and had been reinforced by the Italian Royal Guard. Eugene had decided to continue the retreat to the Adige. The army needed time to recover, and the defenses and stores in that area would allow it to do so. Although Eugene had been defeated by John, he was still concerned about Chasteler. He would trade space for time to meet Chasteler and gain a respite from John. Finally, having lost a battle, Eugene decided to return to Napoleon's plans for a defensive campaign based on the Adige defenses.

Eugene began to recover his nerve. The Viceroy had done poorly up to this point, but defeat strengthened him. Whereas some people are overwhelmed by catastrophe, Eugene managed to reach within himself and find the strength to become an effective commander. His concerns about retreat past the Adige subsided. Talk of returning to the offensive resumed, and the specter of Chasteler began to recede.

From the Piave, the French retired behind the Brenta. Patrols were sent up the river into the Alps, and the information about events in the Tyrol began to change. The revolt in the Tyrol had taken everyone by surprise. Baraguey d'Hilliers had reached Trent on April 14 amid great confusion. There he found the French column that had been ambushed in the Tyrol and took charge of it. With these troops and other reinforcements sent to Trent, Baraguey d'Hilliers had a small corps of 11,300 by April 16.[23] The panic of the French troops in Trent had affected him as well. It was he who had estimated that Chasteler had 20,000 troops, and he had sent this report on to Eugene, which influenced Eugene to hurry and fight at Sacile.

News of Eugene's defeat increased Baraguey d'Hilliers' anxiety. Believing that Chasteler was about to fall on him in overwhelming numbers, he requested permission to retreat down the Adige to Verona. By this time, the Viceroy had regained his composure, and his own patrols reported that Baraguey d'Hilliers was facing at most 5,000–6,000 regular troops and that Chasteler was further north at Brixon or Innsbruck. As Eugene's concern about the Tyrol subsided, his desire to turn against Archduke John grew, and he needed Baraguey d'Hilliers to secure his flank. Attempting to stiffen his nervous commander, Eugene wrote to him stating:

> Even you must admit that the enemy has never vigorously attacked you and that you haven't been disturbed even at the moment of retreating. For the moment, I need to gain time. . . . In case you are forced from your current position by superior enemy forces, you will retreat down both banks of the river, disputing the ground step by step, using every inch of cover that the terrain offers, until you reach Incanale. [In the Rivoli-Chiusa sector] you will employ all measures for defense.[24]

A retreat to Verona was out of the question. Eugene's stern tone had the desired effect. Baraguey d'Hilliers was attacked by partisans and some regular troops. He conducted a fighting retreat, beating off the attacks and eventually reaching Rivoli, where he halted by April 27.

Eugene sent Barbou's division to garrison Venice and then continued his retreat, crossing the Adige by April 27. There he was joined by another infantry division under General Durutte and a cavalry division under General Grouchy. The Army of Italy was now concentrated, and Eugene had 55,500 men along the Adige.[25] John's army, which was approaching the Adige, totaled at most 40,000. The initiative was passing to the French.

Eugene had been considering a return to the offensive for some time. He planned to launch a drive against John's right, turning his flank and driving him against the defenses of Venice. This was a plan of operations according to the Napoleonic practice of envelopment. Before the plan could be put into effect, the army needed some reorganization.

Eugene now had twelve divisions (eight infantry, three cavalry, and the Royal Guard) along the Adige from Rivoli to Venice. The time had come to organize the Army of Italy into corps (it was only when Napoleon learned of the Austrian attack on April 10 that he confirmed Eugene's nominees for corps command). The army was to be divided into three corps and a reserve. General Macdonald joined the army sometime between April 23 and 28 and was given the right or V Corps, consisting of the divisions of

Lamarque and Broussier and a dragoon brigade under Guerin. Grenier was give the center, or VI Corps, consisting of his old division, now temporarily commanded by General of Brigade Abbé,[26] Durutte's division, and four squadrons of the 8th Hussars. Baraguey d'Hilliers had in effect been commanding a corps since his arrival at Trent; his command was designated the left, or XII Corps, and was reorganized to contain the two Italian infantry divisions under Fontanelli and Rusca.[27] The army reserve consisted of Serras' infantry division, three cavalry divisions, and the Royal Guard and was kept under Eugene's personal command.[28] The composition of the reserve would vary as units were sent to reinforce the three corps. In addition, an army artillery reserve was created with expert artillery commander General of Division Jean Sorbier in command. With his arrival, guns were drawn from the infantry divisions and joined with the artillery of the Royal Guard to form the artillery reserve.

The formation of a left, right, center, and reserve for the Army of Italy mirrored on a smaller scale the organization of Napoleon's army, especially in the formation known as the *battalion carré* (see Appendix A), which would allow the Army of Italy to maneuver effectively in any direction.[29] This type of organization reflects the existence of common practices in Napoleonic armies and indicates that methods and procedures were not kept secret at Napoleon's headquarters.

As the Viceroy turned his thoughts to resuming the offensive, he realized that he would have to recross the rivers of northeast Italy. He needed a fast-moving force that could ford the rivers and establish bridgeheads ahead of the main army. To this end, he created a light brigade. Companies of *voltigeurs* (light infantry) were drawn from their parent regiments and combined into three battalions, and a squadron of light cavalry and two guns were added. This brigade was initially placed under the command of General Armand Debruc.

The crisis in Italy abated as Eugene's army reorganized behind the Adige. However, the impact of the defeat at Sacile spread beyond Italy. Napoleon first learned of Eugene's defeat on April 25 after having driven Charles' army over the Danube. The Viceroy was understandably reticent about reporting the events at Sacile, so his original dispatches were deliberately vague. Since Eugene had earlier reported that Lamarque and Pully had arrived before the battle, which they did not, and he subsequently did not mention that they were absent at Sacile, Napoleon was left with the impression that a numerically superior French army had been beaten by the Austrians and that a catastrophe had occurred. What concerned Napoleon the most was that if the Army of Italy had been decisively defeated, John would be free to turn north and threaten Napoleon's flank as he advanced down the Danube toward Vienna. The Emperor's anxiety about the events in Italy is reflected in his letters to Eugene: "I am ignorant about the last battle, the number of men and guns that I have lost, and what caused the defeat. . . . Not having the slightest idea of what took place on the 16th upsets all the calculations for my campaign."[30] Later Napoleon wrote wondering if John's army would, "soon be on my right flank."[31] The uprising in the Tyrol and the threat of John's appearance forced Napoleon to send Marshal Lefebvre and the VII Corps into the Tyrol.

It took a week for a courier to travel between Viceregal and Imperial headquarters. Napoleon's anxiety about Italy intensified from April 25 to April 30. The picture that presented itself was that Eugene was totally incompetent and had failed miserably. By April 30, Napoleon concluded that Eugene would have to be relieved of command and reduced to commanding a corps. He ordered Eugene to write to Marshal Murat, King of Naples, to come and take command of the army.

> I repeat that unless the enemy is already retiring and perhaps in any case, you should write to bring the King of Naples to the army. You will gain merit and glory serving under one older than yourself. You will acquaint him of this fact that you have been authorized by me for this proceeding, and that upon his arrival he will find his letters of command.[32]

By the time Eugene received these orders, the Army of Italy was already advancing and was on the verge of fighting another major battle. The Viceroy never acted on these instructions. Nor would he lose another battle.

CHAPTER SEVEN

Victory in Italy

The bulk of Eugene's army was before Caldiero in the angle between the Adige and the Alpone. The French right was anchored at Arcole; its left extended north behind the Tremegana Creek reaching to Illasi and Cazzano. Other units extended south down the Adige. Archduke John's army faced the French with his main body centered at Villanouva, with units on his left at Montagno and Legnago. On his right, units were positioned around Soave and Monte-Bastia opposite the French at Illasi and Cazzano.

Eugene was completing the reorganization of his army and was planning a counteroffensive. No longer fearing from the Tyrol, he ordered Baraguey d'Hilliers to leave one division to reoccupy Trent and bring his remaining division to Caldiero to join in the offensive against John. The Viceroy had been considering turning John's right, driving him toward Venice while the garrison in that city sortied to the north. If all went well, John's army would be cut off and destroyed. To get in position for the operation, Eugene needed to seize Monte-Bastia. This move would flank John's army and secure a position for an advance into the rear of the Austrian army. The corps of Grenier and Macdonald feinted before Soave and Villanouva while seven battalions from the reserve, spearheaded by the Royal Guard, stormed Monte-Bastia. Realizing the danger to his flank and rear, John counterattacked with eleven battalions the next day, regaining his positions. Eugene marshaled his forces for a major attack on May 2, but by then, the Austrians were gone.

Archduke John first learned of Charles' defeat at Eckmühl on April 27. Orders from Charles arrived two days later. The Austrian commander in chief gave his brother considerable latitude: John could continue his offensive, but he was to hold all of the captured territory and secure the defense of Carinthia and Carniola until Charles could return to the offensive. John was allowed to modify the orders as he saw fit.[1]

John knew that his victorious offensive was over. Eugene's army had grown so strong that it was impossible for him to advance any further. The Austrian commander also realized that his current position would become untenable. As Napoleon drove down the Danube toward Vienna, the likelihood of French forces moving through the Alps to cut off the Austrians in Italy increased. Besides, with Eugene's forces nearing a strength of 50,000, he might not be able to hold his current position if

Eugene attacked. John decided that all he could do was defend the frontiers of the Austrian Empire, provided he got his army out of Italy before disaster struck. Part of his army was to withdraw due east to Carniola to cover the mobilization of the Croatian feudal ban or levy. The rest would ascend the Tagliamento into Carinthia where John could protect the southern approaches to Vienna and draw reinforcements from Hungary.[2]

The withdrawal began shortly after midnight on May 1. Frimont's division, which had served so well as advance guard, was given the role of rear guard. The Austrians headed for Citadella behind the Brenta where they were to rendezvous with the detachment that was covering Venice. To delay the French, the bridges over the Alpone were destroyed.

It took all day for the French to repair a bridge over the Alpone, so the French pursuit did not begin until May 2. Eugene sent Durutte's division south to cross the Adige at Legnago, cross the lower Brenta near Padua, and join with the troops that had sortied from Venice. In addition, Durutte was to cover the movement of supplies from Venice to Eugene's army. The rest of the Army of Italy moved out in pursuit.

The Light Brigade caught up with Frimont at Montebello but was thrown back after a sharp fight. In this combat, General Debruc was wounded and was succeeded in command by Colonel Renaud. The Austrians got behind the Brenta safely, were joined by their forces from the south, and headed for the Piave River after destroying the bridges over the Brenta.

The Viceroy took stock of the situation. The pace of the Austrian retreat would be slowed by the supply wagons that they were loath to lose. In spite of the Austrian reforms, they still could not march as fast as the French. Eugene therefore calculated that he would be able to catch the Austrians just east of the Piave.[3] Much has been made of Napoleon's prediction for the site of the battle with the Austrians in 1800.[4] Although this was not in the same league, Eugene's prediction for the coming battle is a further indication of his improved abilities.

Anticipating a major battle, Eugene ordered a general concentration along the Piave. Durutte was to be reinforced with eight battalions from Venice and rejoin Eugene's army along the Piave, bringing the supply train with him. Realizing he needed a more powerful force to overcome Frimont's rear guard, Eugene increased the Light Brigade to a division. Three more *voltigeur* battalions were created to expand it, two more guns were added, and the cavalry component was increased from two squadrons to a regiment. General of Brigade Joseph Dessaix was given command of this Light Division.

Thinking beyond the forthcoming battle and considering his theater as a whole, Eugene ordered Rusca's division to serve as a flank guard for the Army of Italy. Rusca was to move to Trent, from there march east to Feltre to secure the upper Piave, and serve as a wedge between John and Chasteler. Marmont was informed of Eugene's advance and was ordered to "vigorously attack the enemy" to begin the envelopment called for in Napoleon's offensive operational plan.[5]

The Austrians crossed the Piave on May 6, destroying the bridges behind them as was their custom. During the retreat, John had used a brigade under Joseph von Schmidt to cover the northern flank of his army. Schmidt had been at Bassano on the Brenta and crossed over to the upper Piave. The rest of John's army was some distance to the south near Conegliano. French light forces crossed the Brenta in boats while a bridge over the river was repaired. French light cavalry reached the banks of the Piave on the evening of May 6. The rest of the army arrived at the river the next morning.

The Austrians burned the bridges over the Piave. Without Schmidt's brigade, which was too far away to help directly, John had 30,000 troops massed near Conegliano.[6] With the river behind him and believing he was safe, John halted his army for a day to give it a rest and to allow time for his supply wagons, wounded, and other elements to continue their retreat.

Napoleon's letter telling Eugene to send for Murat to take command of his army reached him on May 6. With over 45,000 troops moving to the Piave, Eugene was confident that he would soon defeat the Austrians. The Viceroy pocketed the letter.

Eugene planned to hurl his army across the Piave to attack and destroy the Austrians. May 7 would be a day of preparation, with the assault launched the following day. Rafts and pontoons were collected for the crossing while the area was reconnoitered.

The Piave flows in a generally northwest to southeast direction from the Alps to the Gulf of Venice; but along the thirteen-mile stretch from Narvese to Ponte-di-Piave, it runs from west to east and it was along this sector that the rival armies faced each other. In this area the river is dotted with many small islands and narrows at certain points to a distance of 350 yards. There were three fords over the river: one at Narvese; a second, four and a half miles downstream at Priula; and a third, two and a half miles further at San Nichiol. The river rose in the afternoon so it could only be forded in the morning. Eugene planned to build a bridge at Priula, the narrowest point of the river, in order to continue the crossing throughout the day.

The terrain beyond the Piave toward Conegliano is flat and is interspaced by dikes, irrigation ditches, and streams. A series of hamlets lie parallel to the river, ranging in distance from a half mile to one and a half miles from the river bank. The most significant of these are Barco, La Mandra, Campana, San Michele, Cimadolmo, and Tezze. A small stream,

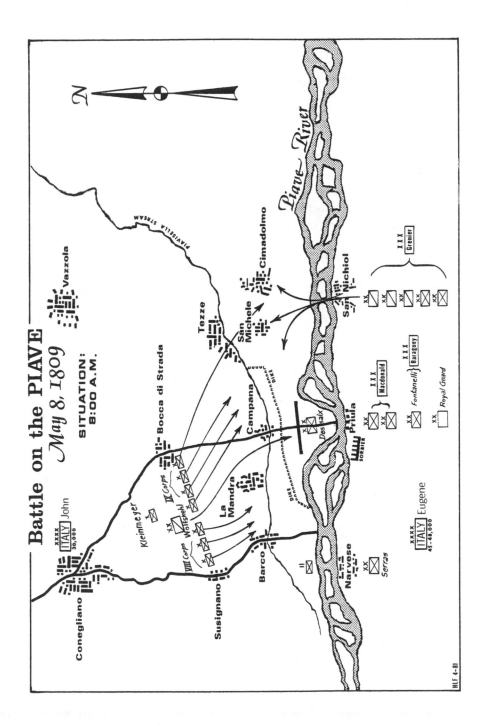

Battle on the PIAVE
May 8, 1809

SITUATION:
8:00 A.M.

Piave River

PIAVISELLA STREAM

Vazzola

Tezze

Bocca di Strada

Conegliano

Kleinmeyer

IX Corps

Wolfskehl

VIII Corps

La Mandra

Campana

Susignano

Barco

Narvese

San Michele

Cimadolmo

San Nichiol

Priula

Dessaix

SORBIER

DIKE

DIKE

ITALY John
30,000

XXXX

Grenier
XXX

Macdonald
XXX

Fontanelli
Baraguey
XXX

Royal Guard

Serras
XX

ITALY Eugene
45-48,000
XXXX

N

HLF 4-81

the Piavisella, runs near or through several of the hamlets. A dike, located a quarter of a mile south of the stream, lies across the southern approaches to La Mandra and Campana. Two primary roads run perpendicular to the river: one through Barco to Conegliano via Susignano; the other from Priula to Conegliano via Campana and Bocca di Strada. The last town, about three miles from Priula, lies halfway between Priula and Conegliano.

On May 7, Eugene sent the 8th Chasseurs across the ford at San Nichiol to reconnoiter. The river bank was lightly held and the French horsemen were able to scout several miles from the Piave. The main body of John's army was sighted between Susignano and Bocca di Strada. The cavalry then returned safely to the French side of the river. If the Austrians remained where they were, Eugene felt confident that he could cross the river the following day.

Eugene planned to have Serras make a feint at Narvese, the ford closest to the Austrian lines, to keep the Austrians occupied while the army actually crossed at Priula and San Nichiol. The army artillery reserve would cover the crossing at Priula, where a bridge would be built. The initial crossing there would be made by Dessaix's Light Division. A bend in the river formed a small tongue of land on the left bank, and Dessaix could anchor his flanks at the base of the tongue to secure the bridgehead, cover the crossing of Macdonald and Baraguey d'Hilliers, and protect the building of the bridge. The ford at San Nichiol was the farthest from the Austrian lines. Grenier's corps, which was reinforced by three cavalry divisions, would cross there. With the cavalry divisions, Grenier was expected to turn the Austrian left flank while Macdonald and Baraguey d'Hilliers engaged the Austrians from the front.

Evidently, Archduke John considered himself safe. His only response to the French reconnaissance on May 7 was to send three infantry battalions to defend each of the three fords. John was more concerned with moving his army through Conegliano and on to the Livenza. Before the French could cross the Piave in force, John expected to be on his way.

The attack began at dawn. Sorbier's massed artillery opened fire on the Austrian battalion on the Priula peninsula. The fire drove the Austrians back, and Dessaix's light infantry waded across the river. Meanwhile, Serras' artillery began firing at Narvese, as did Grenier's at San Nichiol, where the French cavalry began to cross the river.

This broad offensive took the Austrians by surprise. John originally did not plan to dispute the river line. However, he realized that if he did not counterattack, Eugene's whole army could be across the river by noon. The Austrians would not be able to withdraw with all their wagons in time, and John would be forced to fight a numerically superior opponent on open ground. Better to counterattack and wait for the Piave to rise. A portion of the French army could be hemmed in, if not destroyed. However, the suddenness of the French attack rendered the Austrian response disjointed and forced a piecemeal advance against the French. John sent a cavalry regiment and Kalnassy's infantry brigade to the San Nichiol ford. Five cavalry regiments, concentrated into a division under Feldmarschall-Leutnant Wolfskehl, and a battery of twenty-four guns under Generalmajor Reisner were sent to Priula.

It took an hour for Dessaix's division to cross the river and deploy at the base of the tongue of land. Meanwhile, Eugene's engineers began the construction of a pontoon bridge. At this point, the Austrians arrived. The Austrian guns deployed at a range of 800 yards and opened fire against Dessaix's troops. After a brief cannonade, Wolfskehl's cavalry charged. Dessaix's infantry formed into squares and beat off the attack. The Austrians withdrew and re-formed behind their artillery, which resumed its bombardment.

At the San Nichiol ford, the divisions of Sahuc and Pully had safely crossed the river when the Austrian forces arrived. The two French cavalry divisions charged the lone Austrian cavalry regiment, routing it and forcing Kalnassy's infantry to take refuge in Cimadolmo. With the bridgehead secure, Sahuc and Pully waited until Grouchy's cavalry division crossed. Once over, Grouchy remained to cover the crossing of Grenier's infantry while Sahuc's and Pully's divisions galloped off to reinforce Dessaix.

The Light Division was still under pressure, and Eugene wanted to expand the bridgehead. The artillery from Lamarque's and Broussier's divisions, totaling twenty guns, was sent across the river to reinforce Dessaix. Those guns were joined with the four in Dessaix's division and deployed in a massed battery in front of the light infantry. The French guns opened up an artillery duel with Reisner's artillery.

By 10:00 A.M., the cavalry divisions of Sahuc and Pully had joined Dessaix. Eugene was at the Priula bridgehead as well, and having an effective mobile force, the Viceroy decided to attack. At 10:30, with the opposing guns heavily engaged, Sahuc and Pully charged from the flanks. Sweeping around the Austrian gun line, the French horsemen hit Wolfskehl's cavalry from two sides. In a wild melee, Wolfskehl was killed in a saber duel with a French dragoon, and his second in command, Generalmajor Hager, was captured. Leaderless and hemmed in by superior forces, the Austrian cavalry broke and ran. The French cavalry stormed into the rear of the Austrian artillery, killing Reisner and capturing fourteen guns.

Sahuc and Pully re-formed their divisions and set out in pursuit of the Austrian cavalry. Advancing toward Campana and La Mandra, they encountered the infantry masses of the VIII and IX Corps to their front and on their flanks. Wisely deciding not to charge the infantry without support, Sahuc and Pully fell back toward the French infantry.

The destruction of the Austrian cavalry was a major blow. John was reluctant to attack the French in the presence of their superior horsemen. Instead, he deployed his infantry in defensive positions to seal off the French bridgehead, evidently hoping to hold off the French and withdraw during the night. To this end, he deployed most of his infantry behind the dike. The VIII Corps with three brigades went to the right of the Conegliano-Priula road, and the IX Corps with three brigades moved to the left of it. Kalnassy's brigade held Cimadolmo and San Michele, and Kleinmayer's grenadier brigade was kept in central reserve. Unfortunately, this deployment severely extended John's battle line. There were not enough troops to secure the front effectively, and with only a brigade in reserve, there wasn't much left to plug any holes. John's plan of delay might succeed if the French were unable to mount an offensive. With the river rising, this was a possibility.

The Piave had been rising throughout the day and so had the speed of the current. By mid-afternoon, the fords could no longer be used, and the swift current had prevented the completion of the pontoon bridge. The rafts and pontoons destined for the bridge were used to ferry troops and supplies. By 3:00 P.M., the current had become so strong that the ferries had to stop. Nothing could cross the river until after midnight when the level of the river fell. Eugene would have to make do with the forces already across.

By this time, the divisions of Lamarque, Broussier, Abbé, and several battalions of Durutte's division had crossed to join the forces of Dessaix, Sahuc, Pully, and Grouchy. Eugene had 27,000–30,000 men in the bridgehead[7] and was not going to wait for reinforcements. Perceiving the extended nature of the Austrian positions, Eugene decided to attack. He would keep the Austrians engaged in front and turn their weakly held left. Once the Austrian line cracked, massed reserves would finish the job. This was the usual Napoleonic practice.

Dessaix and Sahuc's divisions would hold the left and engage the VIII Corps. The center, opposite IX Corps, was formed by Macdonald's forces, consisting of the massed artillery and the divisions of Lamarque, Broussier, and the regiment from Durutte's division—this would serve as the *masse de rupture* (see Appendix A). The turning movement on the right was to be made by Grenier's corps with Abbé's infantry division and the two dragoon divisions of Pully and Grouchy.

The attack began at 4:00 P.M. Abbé's division attacked Kalnassy's brigade at Cimadolmo and San Michele. The Austrians put up a desperate fight but were outnumbered two to one. They were driven back into Tezze having sustained 1,200 casualties.[8]

With John's left falling back, the time had come to unleash Macdonald's forces. Twenty-four guns opened up a concentrated bombardment against the center of the IX Corps. Perceiving disruption in the enemy ranks, Eugene ordered Macdonald to charge with the bayonet, and the French smashed through the front line of the Austrian troops. Trying to close the breach, John committed his last reserve, Kleinmeyer's grenadiers. The Austrian brigade counterattacked but was hurled back by superior numbers. The IX Corps fell back behind the Piavisella, and the VIII Corps was compelled to conform to the retreat and fell back behind the stream as well.

Sensing victory, Eugene pressed the attack. Dessaix and Sahuc stormed Barco and drove up the road to Susignano. Macdonald took Campana and advanced on Bocca di Strada. Grenier's advance was slowed by irrigation ditches around Tezze, but he eventually captured the village. He then unleashed his cavalry, which swept north to Bocca di Strada. John's army was broken by this final attack, and retreating units streamed into Conegliano. The French pursuit was halted by nightfall, having reached a line from Susignano to Vazzola.

The Austrians, in great disorder, retreated through the night toward the Livenza. They reached Sacile at daybreak, and it was then that some order was reestablished. Frimont was told to organize a rear guard and hold Sacile while the rest of the army continued its retreat to the Tagliamento.

The battle on the Piave was a grave loss to the Army of Inner-Austria. It sustained 5,000 casualties on the first day, and the French picked up 2,000 more prisoners on the succeeding days as a direct result of the battle.[9] In addition, the French captured fourteen guns, thirty caissons, and a great number of other vehicles and horses, all of which were used to move and support Eugene's army. French casualties have been estimated between 700 and 2,000.[10] Considering the length of combat and the strength of the rival force, the latter figure is probably closer to the mark.

The pursuit of John's army began at dawn. Grouchy was given command of all three cavalry divisions and ordered to sweep the Italian plain. Grenier was to follow with Dessaix and Abbé. Macdonald's corps remained at Conegliano while the bridge at Priula was completed so the rest of the army could cross.

Eugene planned to take his army toward Laibach in accordance with Napoleon's offensive operational plan. However, new orders from the Emperor arrived on May 9. Napoleon's letter, dated May 1, told of his own drive down the Danube. He predicted that his victory at Eckmühl and advance on Vienna would force a retrograde movement by John. The Viceroy was ordered to pursue John, invade Carinthia via Villach, and effect a junction with Napoleon's army at Bruck.[11] Napoleon wanted to use the Army of Italy to secure his flank and rear areas while he took his own army north of the Danube to engage Archduke Charles. The major invasion routes into Carinthia lay between the Tagliamento and Isonzo river valleys, so Eugene could continue on his present course until reaching the Tagliamento.

Grouchy and Dessaix caught up with Frimont's rear guard at Sacile late on the afternoon of May 9. The Austrians were driven from the town and continued their retreat to Valvasone. From there, Frimont turned northeast, crossed the Tagliamento at Spilembergo, and followed the route of John's army, joining it at Saint Daniel on May 10.

The speed of Eugene's pursuit cut off what remained of Kalnassy's brigade from John's army. Kalnassy had been driven south away from John's army during the battle on the Piave. He tried to rejoin the main body via Pordenone or Valvasone, but Grouchy's cavalry corps always got there first. Despairing of reaching John, Kalnassy continued east toward the Isonzo and Carniola.

Eugene's spearhead crossed the Tagliamento at Spilembergo on May 11. From there, the Viceroy ordered Grouchy to continue eastward to Udine while Dessaix followed Frimont. The Light Division caught up with Frimont's rear guard at Saint Daniel on the afternoon of May 11 (John had already left with the main body). Frimont was in a defensive position with 4,000 men. Under Eugene's command, Dessaix executed a double envelopment and severely mauled Frimont's division. His command sustained 1,900–2,100 casualties; the French, 200–800.[12]

Frimont rallied his forces at Venzone but was pushed out of that town by Dessaix the following morning. Frimont picked up the Austrian force that was blockading Osoppo and followed John's army up the river valley of the Fella, destroying the bridges behind him. In all, the Austrian forces that recrossed the Carinthian frontier numbered 18,000–19,000,[13] less than half the number that had crossed the frontier a month before. The only Austrian forces remaining in Italy were Zach's brigade (which was blockading Palmanova) and Kalnassy's. Together they crossed the Isonzo into Carniola on May 13, and Italy was free of invaders.

The Italian campaign was an unmitigated disaster for Austria. With smaller numbers than the French, John had no chance to conquer Venetia, let alone Lombardy. The subsidiary invasions of the Tyrol and Dalmatia served only to weaken the thrust into Italy. It was a miracle that John's army got all the way to the Adige. Had it not been for Eugene's poor generalship, John's victory at Sacile would not have occurred. Eugene had sufficient forces to defend the Livenza until reinforcements arrived. At that point, the French would have been strong enough to attack effectively.

It is evident from the performance of the Army of Inner-Austria that the prewar reforms had done wonders. The tactical employment of Austrian troops in the battles in Italy showed a high degree of coordination and resilience. The stout fighting abilities of Frimont's division is particularly noteworthy. It was outmaneuvered at Saint Daniel, but after that battle it was re-formed and remained in existence in spite of the heavy casualties it suffered throughout the campaign. The Austrian army fought well on the Piave. John's belief that they were safe from attack put his army in danger. Had his army been closer to the river, the battle may have gone the other way.

The losses sustained by the Austrian army on the Piave were grave. However, the Piave was not a miniature version of Austerlitz, where losses totaled 7,000 at most. John lost half his army during the entire campaign. It was not just the losses on the Piave that cost John his army; rather it was the cumulative cost of the campaign and the series of battles fought since the start of the war. Comparing the losses for the rival forces at Sacile and the Piave, the two battles canceled each other out, as Eugene's losses at Sacile almost matched John's at the Piave. The same symmetry can be observed at the combats at Pordenone, Monte-Bastia, and Saint Daniel. The French lost at Pordenone, drew at Monte-Bastia, and won at Saint Daniel. The campaign in Italy was determined by the cumulative effect of all of these battles combined with the fact that the French had the bigger army from the start. The Austrian reforms closed the qualitative gap between them and the French: the Austrians could now compete with the French one for one. In such an attritional situation, the Austrians were bound to lose; Eugene could take the losses better than could John.

Although the invasion was a strategic mistake, the Austrian planning had a distinctly modern tone. A surprise assault was launched on a strategic front extending from Dalmatia to Bavaria. Coordinating partisan revolts to disrupt the French rear areas while attacking them with regular forces from the front displays a definite operational sophistication. Unfortunately for the Austrians, the revolt fizzled in Italy. However, the impact of the Tyrolean revolt on operations in Italy and Germany indicated how powerful was the combination of partisans and regular forces.

The Italian campaign tore the heart out of John's army. The best-trained troops and officers sustained the most casualties. Although the numbers could be made up, the quality and training could not. One must consider the course of this war had the Austrians chosen not to invade Italy but to use John's army to defend the southern frontiers while Charles attacked on the Danube. In spite of his larger army, Eugene would have been hard pressed to breach the southern defenses and the Austrian Alps if John had an effective and well trained army to bar the way. The French would have been smashed trying to break through, and the course of the war could have been changed; this aspect will be examined later in further detail.

What of the French performance in this campaign? Napoleon was responsible for the incomplete nature of the army's structure at the start of the campaign. Had the corps been in place and the army concentrated, the defeat at Sacile may not have occurred. Napoleon's offensive and defensive operational plans served to guide Eugene throughout most of the campaign, with the exception of Sacile. Those plans incorporated Napoleon's operational methods of war. If the first week of operations showed Eugene at his worst as a commander, his subsequent actions showed him at his best. He regained his nerve and became an effective commander, as his counteroffensive in Italy has shown. What is more important is that his approach to war and the conduct of his army illustrates a broad understanding of Napoleonic methods and practices. Eugene's stillborn scheme to turn John's right and drive him toward Venice was an example of the *manoeuvre sur les derrières*. The division of the Army of Italy into a left, center, right, and reserve mirrors Napoleon's organization of his army in 1806 and on the Danube in 1809. His conduct of the battle on the Piave exhibited an understanding of the *masse primaire, masse de manoeuvre*, and *masse de rupture* as Napoleon used them (see Appendix A). The similarities in the conduct of operations between armies commanded by Napoleon in person and those of this Army of Italy indicate that Napoleon's methods had been inculcated throughout the French army. Eugene was no Napoleon, but these similarities show that there was commonality in approach to organization, operations, and tactics. Such accepted practices throughout a military institution are now called *doctrine*. This commonality cannot guarantee victory, but it does provide the basis that makes victory possible, and that is what we see in the case of this Italian campaign.

CHAPTER EIGHT

The March on Vienna and the Battle of Aspern-Essling

The strategic situation in the Danubian theater of operations on April 26 was as follows: the main Austrian army under Archduke Charles had been defeated but not destroyed. Two thirds of Charles' army was on the north bank of the river while one third, under Hiller, was located south of the river. The Tyrol was in revolt, and the divisions of Chasteler and Jellachich were in position to operate against Napoleon's operational right. The Emperor had learned of Eugene's defeat at Sacile and subsequent retreat to the Adige. The almost hysterical tone of some of Napoleon's dispatches[1] indicates he feared John would take the Austrian army north from Italy to march against his flank. Napoleon was well aware of the interrelationship between the two theaters. The Danubian theater was the primary one, and Napoleon expected that any offensive movement he made on the Danube would have some impact on Italy. John would be compelled to respond to Napoleon's moves in one of two ways: he could attack Napoleon's right and his line of communications, or he could retreat in an effort to protect Vienna and the heart of the Danubian monarchy from invasion. The extent of John's ability to intervene on the Danube would depend on how well Eugene could engage John's forces. On April 26, the ability of Eugene's army to do that was unclear, which is why Napoleon considered relieving Eugene of his command.

Napoleon could not wait to get a clear picture of Italy in relation to the operations of his own army. It would be a gamble, but he wanted to finish Charles' forces in spite of the threat from the south. Napoleon's center of gravity was below the Danube facing just a third of Charles' army. The good roads south of the Danube offered an easy march to the Austrian capital of Vienna. An offensive below the right bank of the Danube to Vienna had always been part of Napoleon's operational plans. The first phase of his campaign plan had been to stop the Austrians and wrest the initiative from them; that phase had been accomplished, and it was now time to move to the next offensive phase. Napoleon reasoned that by marching on Vienna he might make the Austrian Emperor sue for peace. If not, he could drive a wedge into the heart of the Austrian Empire, keep the armies of Charles and John divided, and maneuver between the two, defeating each in detail—a strategy based on interior lines.

To this end, Napoleon recalled Davout's III Corps that had initially been

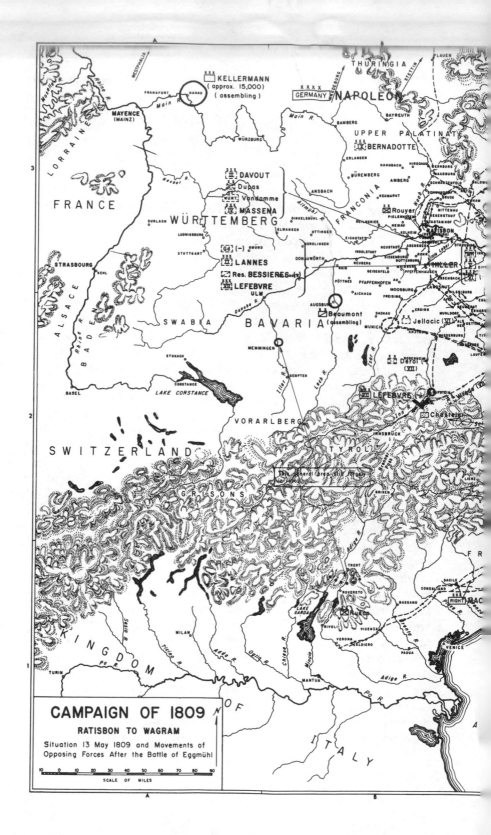

CAMPAIGN OF 1809

RATISBON TO WAGRAM

Situation 13 May 1809 and Movements of
Opposing Forces After the Battle of Eggmühl

SCALE OF MILES

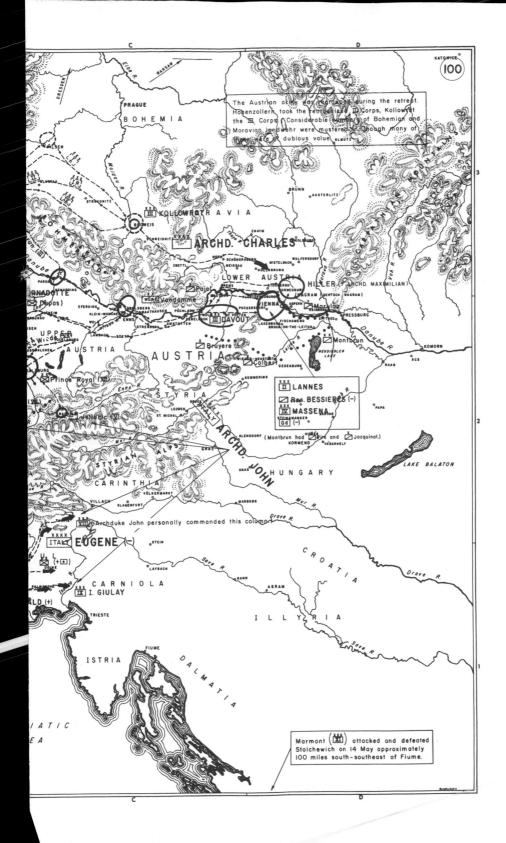

sent north of the Danube to pursue Charles to the south bank. The advance on Vienna would be made by Bessières, now commanding the Cavalry Reserve Corps; Massena's IV Corps; Vandamme's VIII Corps; Davout's III Corps; and the II Corps, which had been reorganized and placed under Marshal Lannes.[2] To protect Napoleon's lines of communication, Bernadotte's IX Corps was ordered to march from Dresden to Ratisbon. Lefebvre's VII Corps was sent toward Munich to deal with Jellachich and the Tyrolean revolt. Lefebvre would later move to the Tyrol to crush the rebellion, deal with Chasteler, and secure Napoleon's operational right. Moving up to reinforce the army was the Imperial Guard, consisting of two infantry divisions and a cavalry division. Napoleon had earlier organized a X Corps consisting of Wesphalian troops under his brother Jerome, the King of Westphalia. To protect his operational left from any incursion from Bohemia and to check any possible revolt in northern Germany, Jerome was ordered to send units to Dresden. In addition, a new Corps of Observation under Marshal Kellermann was organized and deployed at Sedan, Metz, and Mayance. Kellermann would support Jerome if a threat developed from Bohemia or in northern Germany.

Davout, Lannes, Massena, and Bessières were sent out to pursue Hiller, whom Napoleon hoped to catch and destroy. Two columns of pursuit were organized: the northern column consisting of the III, IV and VIII Corps went by way of Passau along the southern bank while the II Corps and the bulk of Bessières' command went further south via Burghausen.

The Danube is a mighty, dendritic river and runs from west to east. The feeder rivers, in particular the Inn, Traun, and Enns, all run from south to north and serve as obstacles for any force moving down the Danube on the southern shore. The retreating Austrians were able to delay the French by destroying the bridges over the rivers and by fighting rearguard actions. Hiller somewhat enjoyed being separated from Charles and having the opportunity for independent command. The division of the main Austrian army gave Emperor Francis an opportunity to intervene more actively in the campaign by giving direct orders to Hiller. Francis wanted Hiller at least to delay the French; if the French could be kept from the capital, then all the better. Hiller was crossing the Traun on May 3 when Massena and Bessières caught up with him. Hiller had 22,000 men and seventy guns under his command in and near the town of Ebelsberg, with 8,000 more in supporting distance.[3] The town commanded the bridge, over the Traun. Behind the town was a ridge. The bulk of Hiller's forces were on the ridge but some units were in the town and his rear guard was still on the west bank.

The French promptly attacked and broke through the Austrian rear guard. The French and Austrian units became intermixed as both sides tried to get over the bridge. French troops of Louis Coehorn's brigade of

Michel Claparede's division managed to get into the town. The brigade was later followed by the rest of Claparede's division. The Austrians responded, but the counterattack was piecemeal and badly coordinated, and Hiller failed to commit the bulk of his forces. The key to controlling Ebelsberg was the castle located within the town, and a seesaw fight for it ensued. Claparede was reinforced by Claude Legrand's division. Finally the French stormed the castle. A Badenese brigade reinforced the French. Hiller now opened up a heavy artillery bombardment from the heights above the town. Ebelsberg was soon burning, but Hiller decided to abandon any further attempts to retake it and ordered a retreat. Meanwhile, Lannes had secured a crossing over the Traun further downstream at Lambach. Hiller continued to retreat, burning bridges behind him until he reached Mautern.

The French won the fight at Ebelsberg. At first glance, this appears to follow the usual pattern of French victories over Austrians. However, probing deeper, we notice the high proportion of Austrian guns and their use, an indication of the potential firepower that could be produced by the Austrian corps. It was a warning to the French that went unheeded.

Hiller was ordered by Charles to cross to the north bank of the Danube in order to unite with him. With part of his forces Hiller crossed over to the north bank at Mautern on May 7, leaving 10,000 troops on the south bank to cover the approaches to Vienna. In Italy on the following day, Eugene attacked and defeated Archduke John on the Piave.

Napoleon ordered Davout to establish a strong bridgehead over the Danube at Linz while the rest of the army marched on Vienna. Hiller meanwhile tried to get to Vienna before Napoleon. The French won the race, reaching the outskirts of the capital on May 10.

Vienna was garrisoned by 35,000 poorly trained troops under Archduke Maximilian. Neither the Austrian commander nor his troops had much stomach for a fight. Napoleon began to bombard the city on May 11, and Maximilian decided to evacuate that night. The Austrians marched over the Danube on May 12, burning the bridges behind them. The French took possession of the city the next day, capturing great stores of food, ammunition, arms including 100 guns, and gold totaling 4,500,000 florins.

Napoleon had taken Vienna and driven a great wedge into the heart of the Austrian Empire. Yet no peace followed. Charles' army was still in the field and had to be dealt with. Napoleon's eyes turned to the north. What had Archduke Charles been doing?

Archduke Charles had gotten two-thirds of his army across the Danube by April 24. It was a trying time for Charles. Everything he had worked for seemed to be lost. The downward spiral in Austria's fortunes since the failure of the Prussian alliance continued. The army that he raised had been beaten. Instead of a victorious offensive into Germany, Charles had to look to the defense of the Danubian monarchy. Although only half of his army had been effectively engaged in the April battles, Charles believed that the war was already lost. A severely shaken man, he sent a letter to Napoleon attempting to negotiate peace on April 28. The Emperor ignored the letter; either he believed the message to be a trick or he wanted the Austrians to suffer further humiliation. Members of the imperial Austrian court, including Empress Ludovica, were enraged by the course of events and attempted to remove Charles from command. But as his biographer Rothenberg mentions, "There was no one to replace him."[4] What was to be done?

Charles failed to intellectually grasp and master the type of warfare that he had just fought, and the fast-moving and uncertain operations fought in April had completely unnerved him. He could not effectively direct his far-flung corps, and he disliked the pressure of mobile operations. Unlike anything he had tried to do before, it was too new and too frightening. Nor did it seem to Charles that his corps commanders could operate in modern war. They failed to show enough initiative, and staff work at all levels was poorly done. Austrian command and control, so vital in fast-moving operations and distributed maneuver, could not match the French. Believing that his army could not fight on a broad front or conduct mobile operations, Charles fell back upon centralized command and control. He declared that the army would no longer be deployed on broad fronts and fight in a distributed manner, relying on decentralized command and control and the initiative of corps commanders. Instead, the army would fight and maneuver as a single strategic block. Mobile operations were abandoned in favor of positional warfare, which would offer the least risk. To this end, the army would be reorganized and the corps system abolished. An attempt to return to the safe and secure methods of eighteenth-century warfare, this was an admission that the army could not conduct modern nineteenth-century operations.[5]

But what Charles was attempting to do was impossible in the current situation. One could move armies as single blocks if the armies numbered about 50,000, as they often did in the eighteenth century and as Wellington did in the Peninsula. However, Charles' immediate field army numbered well over 100,000, and Charles was responsible for the movement and coordination of all of Austria's field armies fighting on a strategic front hundreds of miles long. It was physically and strategically impossible to keep so large an army concentrated as a single block for an extended period of time, just as it was impossible to abolish so vital a structure of command and control as the army corps system. Charles himself realized, to a point, that his goals were illogical. Even though he had decreed that army corps were abolished, the corps system was retained.[6] There was no other way to control so large an army.

Charles had moved his army north into Bohemia after crossing at Ratisbon. Davout had originally been sent to pursue him, but he had been recalled to the southern bank of the Danube. Consequently, the French lost sight and knowledge of the exact location of the Austrian main army. The Bohemian mountains served to give the army some refuge and to screen them from the French, but the poor roads in Bohemia ensured that the French would get to Vienna before them. This respite gave the Austrians some time to regain their morale and cohesion. By May 16, Charles' army was approaching Vienna from the north and had rejoined Hiller. Napoleon, however, did not know that Charles and his army were now within striking distance of Vienna.

Charles reorganized his army. Earlier, on May 2, he had decided that he needed a force to defend Bohemia from a possible French attack from Saxony, so he had organized a new corps to watch the Saxon frontier. This new force was designated the II Corps, and the old II Corps was renumbered as the III Corps.

On May 16, after the junction with Hiller, the I and II Reserve Corps were amalgamated into a single I Reserve Corps consisting of two grenadier divisions and two cavalry divisions, all under the command of Prince Lichtenstein. This more closely followed the French practice; the French reserves consisted of the Cavalry Reserve Corps and the Imperial Guard. An Army Advance Guard under Feldmarschall-Leutnant Klenau was organized to cover the front of the army in the hope of providing an effective screen for the army's movements and to provide better reconnaissance to Archduke Charles.

The strategic situation by the middle of May was interesting. Archduke John had been defeated in Italy, and Eugene's forces were invading southern Austria and Croatia. Napoleon had driven a wedge 200 miles long into the heart of the Austrian Empire; the base of the wedge was on the Inn and the apex was at Vienna. However, Napoleon's operational fronts faced north and south. To the south was the Tyrolean insurrection and the forces of Chasteler, Jellachich, and Archduke John. Napoleon had to leave Lefebvre, with Bernadotte and Vandamme in support, to guard the southern front and to protect the French lines of communication. Additional Bavarian units were needed to secure Bavaria from Tyrolean raids.[7] To the north was Charles' army, which posed the greatest problem for Napoleon. The front ran parallel to Napoleon's line of communication. Charles had alternate lines of communication running perpendicular to the front toward Prague and Brunn or parallel to the front toward Budapest, making Napoleon's communications more vulnerable to attack.

The Danube served as a help and a hindrance. An asset to the side that wished to use it as a screen to hide its movements, it was a liability when it masked the movements of the opposing army and served as an obstacle for an offensive across the river from either side. From a theoretical point of view, Charles had the advantage. He could use the river to conceal his movements and attempt to cross the river into the French rear to threaten their lines of communication. In short, he could conduct a *manoeuvre sur les derrières* against Napoleon. Such a maneuver would compel the French to evacuate Vienna and would force them to fight a battle to re-open their communications on ground of the Austrians' choosing.

That is what Napoleon would have done if the situation were reversed, but Charles was not Napoleon. Charles made a half-hearted attempt to ac-complish this by having Kollowrat attack Bernadotte at the Linz bridge-head on May 17, but the Austrians were beaten back. It would take bold-ness and rapidity in offensive maneuver for Charles to execute a major blow against the rear of the French army, and it also would require a series of feints on a broad front to keep the French guessing where the main blow would fall. To accomplish this would require distributed maneuver on a broad front, the exact type of operation that Charles had just decided to avoid. His temperament would not allow him to direct that type of op-eration, nor did he believe that his army was up to it, and the poor show-ing of corps commanders and staff may have indicated that he was right. One can speculate about whether a bolder and more imaginative com-mander might have been more aggressive. But there were no better re-placements, and there seemed to be little talent among the corps com-manders, judging from their performance thus far. Charles himself had suffered from great pessimism since the start of the war, and whether his evaluation of the abilities of his army was based on a realistic appraisal or his own pessimistic personality cannot be fully answered.

Napoleon did not know the location of the main Austrian army; that was hidden from him by the Danube. The French Emperor had strategi-cally split the Austrian armies. However, his lines of communication could be interrupted.[8] Even with the occupation of the Austrian capital, Em-peror Francis did not sue for peace. Napoleon believed that the destruc-tion of Charles' army was necessary to win the war. Napoleon would have to cross to the north bank of the Danube and bring Charles to battle, and the Emperor had complete confidence in his ability to do so. The past performance of the Austrian army, which had so disheartened Charles, bred contempt in Napoleon, who thought that once he located the Aus-trian army, it could be easily crushed. But he had to cross the Danube first.

The Danube River was a formidable obstacle. The melting snows and rainstorms of spring produced sudden and dangerous floods. North and east of Vienna the Danube contained numerous islands of varying sizes, and it was necessary to cross one, if not several, of these islands to move from one bank to the other. The existing bridges had all been destroyed. There were Austrian forces on the north bank, and the French did not know how many. Napoleon thought that Charles and the bulk of his army were further away at Brunn and that a crossing could be made before Charles could intervene.

Napoleon had Lannes' II Corps, Davout's III Corps, Massena's IV Corps, Bessières' Cavalry Reserve Corps, and the Imperial Guard available for operations. A portion of the Guard, the elite of Napoleon's army, had arrived by this time (other units of the Guard were in Spain). The Guard in Vienna consisted of two reduced infantry divisions: one division was the Young Guard (six battalions), under Philibert Curial; the other, the Old Guard (four battalions), under Jean Dorsenne. There was also a Guard Cavalry Division of seven and a half squadrons and some Guard artillery, but it was not at full strength.[9] In all, Napoleon had about 95,000 troops and 144 guns in the force he would use to cross the river.[10]

A series of sites to cross the Danube were examined. Napoleon decided not to try to rebuild the old bridges because they were too closely guarded by the Austrians. Other sites were examined and discarded as being too close to the enemy or too far from Vienna. Finally, a place was chosen four miles downstream from Vienna at Kaiser-Ebersdorf. Here the river was wide but of shallow depth, and the current was slower. At this spot the Danube was cut into different arms by several islands, the largest of which was Lobau Island. A bridge could be floated downstream in sections and then anchored to link the south bank to a small island in the river and the small island to Lobau. This would serve as the main bridge. Lobau Island was four miles square and would mask the French crossing from the south bank to the southern shore of Lobau. The northern shore of Lobau was 125 yards from the north bank of the Danube. Once on Lobau, the French would erect a pontoon bridge from Lobau to the north bank at a spot where a small tongue of land was formed by a turn in the river. This tongue became known as the Mühlau salient. A mile north of the salient was the main road that ran along the north bank and a series of small villages, each a mile apart, were situated along the road. The village of Aspern was a mile and a half northwest of the salient, and the village of Essling was a mile and a half northeast. Gross-Enzersdorf was a mile southeast of Essling.

In the original plan, the French were to take precautions. There was to be a feint made at the broken Nussdorf bridge just north of Vienna, and protective screens and palisades were to be built to protect the bridges from being broken by objects floated down from upstream. However, these plans were abandoned. The French command still believed that

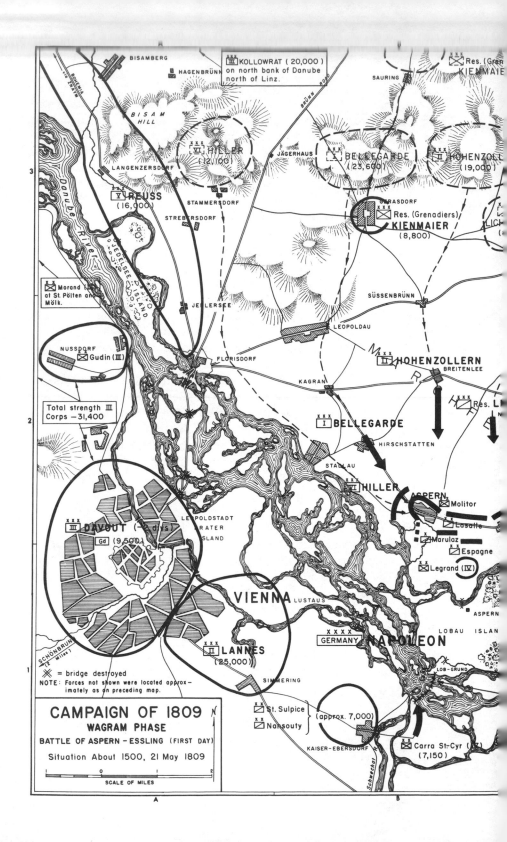

III KOLLOWRAT (20,000)
on north bank of Danube
north of Linz.

BISAMBERG

HAGENBRÜNN

XXX Res. (Gren
KIENMAIER

SAURING

BISAM HILL

VI HILLER
(12,100)

JÄGERHAUS

XXX BELLEGARDE
(23,600)

II HOHENZOLL
(19,000)

LANGENZERSDORF

V REUSS
(16,000)

STAMMERSDORF

STREBERSDORF

GERASDORF

XXX Res. (Grenadiers)
KIENMAIER
(8,800)

LICH

Morand ( )
at St. Pölten and
Mölk.

JEDLERSEE

SÜSSENBRÜNN

LEOPOLDAU

NUSSDORF
Gudin (III)

FLORISDORF

KAGRAN

II HOHENZOLLERN
BREITENLEE

Total strength III
Corps — 31,400

I BELLEGARDE

HIRSCHSTATTEN

XXX Res. L

STADLAU

VI HILLER

ASPERN
Molitor

III DAVOUT (- 1 div.)
Gd (9,500)

LEOPOLDSTADT
PRATER
ISLAND

Lasalle

Marulaz

Espagne

Legrand (IV)

VIENNA LUSTAUS

ASPERN

LOBAU ISLAN

XXXX
GERMANY NAPOLEON

II LANNES
(25,000)

LOB-GRUND

SIMMERING

= bridge destroyed
NOTE: Forces not shown were located approximately as on preceding map.

CAMPAIGN OF 1809
WAGRAM PHASE
BATTLE OF ASPERN-ESSLING (FIRST DAY)
Situation About 1500, 21 May 1809

St. Sulpice

Nansouty

(approx. 7,000)

KAISER-EBERSDORF

Carra St-Cyr ( )
(7,150)

SCHÖNBRUNN

1   0   1   2
SCALE OF MILES

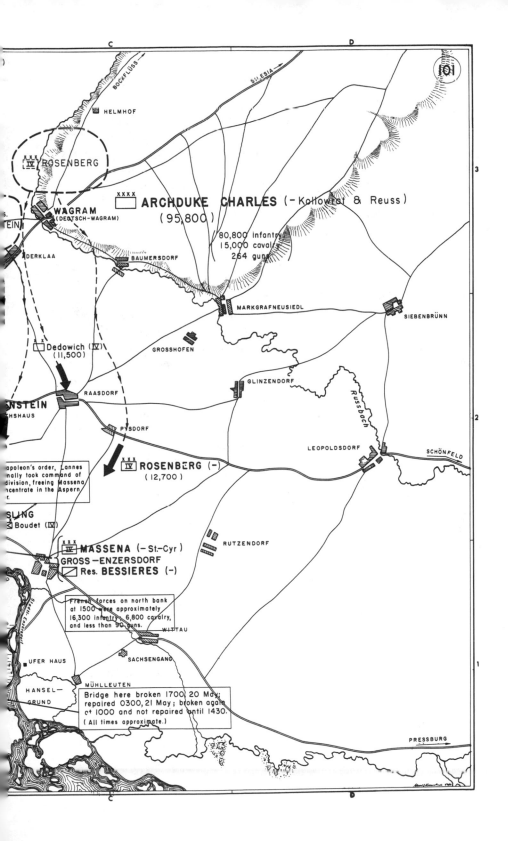

101

BOCKFLÜSS

SILESIA

HELMHOF

xxx
IV ROSENBERG

xxxx
ARCHDUKE CHARLES (- Kollowrat & Reuss)
(95,800)

80,800 infantry
15,000 cavalry
264 guns

WAGRAM
(DEUTSCH-WAGRAM)

STEIN

DERKLAA

BAUMERSDORF

MARKGRAFNEUSIEDL

SIEBENBRÜNN

Dedowich (IV)
(11,500)

GROSSHOFEN

RAASDORF

ENSTEIN

CHSHAUS

GLINZENDORF

Russbach

PYSDORF

LEOPOLDSDORF

SCHÖNFELD

xxx
IV ROSENBERG (-)
(12,700)

apoleon's order, Lannes
nally took command of
division, freeing Massena
oncentrate in the Aspern
r.

SLING

Boudet (IV)

RUTZENDORF

xxx
IV MASSENA (- St.-Cyr)
GROSS—ENZERSDORF
Res. BESSIERES (-)

French forces on north bank
at 1500 were approximately
16,300 infantry, 6,800 cavalry,
and less than 90 guns.

WITTAU

UFER HAUS

SACHSENGANG

HANSEL—
GRUND

MÜHLLEUTEN

Bridge here broken 1700, 20 May;
repaired 0300, 21 May; broken again
at 1000 and not repaired until 1430.
(All times approximate.)

PRESSBURG

Charles was at Brunn and that speed was essential to cross the Danube before he arrived. The crossing of the Danube was to begin on May 18.

On the evening of that date, units of General Molitor's infantry division crossed the Danube to Lobau Island by boat and attacked the Austrian guards there, driving them off. The construction of the main bridge linking the south bank to Lobau began that night. During May 19, Molitor occupied all of Lobau Island and sited a battery of six guns to cover the Mühlau salient. Construction of the main bridge continued throughout that day and into the next. By midday on May 20, the main bridge was completed, and Massena and his IV Corps crossed over to Lobau. Construction of the pontoon bridge linking Lobau to the north bank of the Danube began. Two hundred *voltigeurs* crossed from Lobau to Mühlau by boat to cover the building of the bridge. By 6:00 P.M., a bridge of fifteen pontoons and three trestles was in place, and the IV Corps began to cross to the north bank. The towns of Aspern and Essling were undefended and quickly occupied. Meanwhile, the main bridge was broken by a large hulk sent downriver by the Austrians. The breach delayed the crossing to Lobau of any more troops that night. The bridge was eventually repaired, but this was the first of many breaks to the bridge that would have an effect on operations.

Initial French reconnaissance patrols reported only a light enemy presence. Napoleon was with his commanders at the bridgehead that evening. Looking north, the campfires of a large host could be seen. The Emperor was convinced that this was not Charles' main army. Lannes believed so, too, but Massena demurred. The decision was made to continue the operation, but unfortunately for the French, the campfires were indeed those of Charles' army.

Archduke Charles and the bulk of his army had been near the river since May 16. Charles had an observatory on the Bisamberg Heights and thus an excellent view of French movements on Lobau and the Mühlau salient. As French activities progressed on May 18 and 19, Charles called in reinforcements in preparation for a battle. The terrain immediately north of Lobau Island consisted of a broad flat plain known as the Marchfeld. It was here that the Austrians conducted peacetime maneuvers, so they were very familiar with the ground. There was a series of villages that dotted the plain, including Aspern, Essling, Gross-Enzersdorf, Breitenlee, Sussenbrunn, and Aderklaa. The Marchfeld was ten miles wide and was bordered to the north by the Russbach stream, behind which the ground rose to form a low ridge. On this ridge were the villages of Wagram, Baumersdorf, and Markgrafneusiedl. Six miles to the east of Aspern the ground rose into a series of hills that formed the eastern and northeastern boundaries of the Marchfeld, and that was where the Bisamberg Heights were located.

The situation greatly favored the Austrians. Not only did they know the terrain, but the observation post on the Bisamberg Heights provided Charles with something every commander dreams of but rarely gets: perfect intelligence about the location, movements, and intentions of the enemy forces. Charles had a comfortable superiority in numbers, his forces totaling 133,611 troops[11] and almost 300 guns. Charles' army was concentrated, so there would be no need for distributed maneuver. Unlike the operations in April, the battlefront would be narrow, reducing the need for independent action on the part of the corps commanders and allowing Charles to more easily direct all of his forces. The flat and open nature of the terrain would minimize the French advantage in light infantry tactics, which had characterized the battles of April. Instead, the ground would enable large masses of artillery to be deployed, and here the Austrians had a distinct superiority. The Austrian corps structure would more easily allow the corps commanders to coordinate the three combat arms and in particular would allow them to get their guns into action. Finally, in spite of the defeats in April, the Austrian command was gaining in experience.

On the evening of May 20, Charles' forces were deployed as follows: Kollowrat with the III Corps (20,000) was north of Linz observing the French bridgehead there. From Bisamberg to Muhlleuthen, a front sixteen miles in length along the north bank of the Danube, was the army's advance guard under Klenau (5,905 troops).[12] This was a screen. The main body of the army was situated among the Bisamberg Heights and behind the Russbach at Wagram. Charles left his V Corps under Prince Reuss (16,000) deployed along the north bank of the Danube upriver from Vienna to cover his right flank and rear. The rest of the army, totaling 95,800 troops and 264 guns,[13] would attack the French the following day.

By midday on May 21 the French had three infantry divisions and some cavalry in the bridgehead. Essling was held by Jean Boudet's division, which was placed under the command of Marshal Lannes. Aspern was held by Massena with the divisions of Molitor and Legrand. The cavalry divisions under Marshal Bessières held the center between the two villages. Aspern, the larger of the two hamlets, had several streets, a church, and a cemetery; Essling was dominated by a large granary. Both villages consisted of strong masonry buildings that would serve as defensive strongpoints, and both were surrounded by dikes. The two villages were a mile apart and were linked by a slightly embanked road. Immediately south of Aspern was a small wooded island called Gemeinde-Au, which closed the flank from Aspern to the Danube. Essling was more exposed to an attack from the east and southeast. The two villages would serve as the anchors of any defensive line.

Charles planned to attack the French bridgehead with five converging columns. The first three, consisting of Hiller's VI Corps, Bellegarde's I Corps, and Hohenzollern's II Corps, formed the Austrian right and would attack Aspern. Rosenberg's IV Corps formed the left and was divided into two columns. This wing was to attack Essling. The cavalry divisions of the Reserve Corps under Lichtenstein formed the Austrian center and were to fill the space between the two villages. The grenadier divisions would be held in reserve under Kienmayer. Charles called upon his troops to make a supreme effort. In a proclamation to them he declared, "This decisive battle will be waged under the eyes of our Emperor and of the enslaved inhabitants of our capital, who look for their enfranchisement to the bravery of the army."[14] At 10:00 A.M., Charles gave the order for his army to attack, and the white-coated masses of the Austrian army advanced with regimental bands playing.

The French were taken by surprise. To make matters worse, the main bridge was broken again at midday by a series of floating Austrian missiles. However, the distance from the Austrian line of departure to the French front lines was about six miles, and it took several hours for the Austrian host to come up to the French positions. Moreover, the front contracted as the Austrians advanced with the result that there was not enough room for so huge a force to deploy effectively. Consequently, the Austrian columns bunched up and were compelled to come up to the French piecemeal.

The advance units of the Austrian I Corps attacked Molitor's outposts at 1:00 P.M. Molitor pushed them back, then took up strong defensive positions in Aspern. The Austrians launched a series of attacks against Aspern, but the strong buildings, the piecemeal nature of the Austrian attacks, and superb combat leadership enabled Molitor to parry them time and again.

However, the pressure against Aspern was mounting. By 4:00 P.M. three Austrian corps were ready to attack Aspern. As a preliminary, the Austrians used their artillery in a tremendous bombardment. "The Austrian artillery was at this time pouring from its largely superior number of guns and howitzers a perfect deluge of shot and shell upon the defenders of Aspern."[15]

Also at 4:00 P.M., to take pressure off Aspern, Napoleon ordered his cavalry to charge. Marulaz's light cavalry division charged into Hohenzollern's masses, but the Austrian infantry coolly formed squares to receive them. The French horsemen swirled around the squares, taking heavy casualties, and then withdrew. At the same time, Marshal Bessières sent in Lasalle's light cavalry and Espagne's heavy cavalry division against Lichtenstein's cavalry. There was charge and countercharge between the villages, the French trying desperately to break through. General Espagne was killed in a saber duel and the French cavalry eventually withdrew to a position between Aspern and Essling.

By 5:00 P.M., the Austrian IV, I, and II Corps were finally in position for a major attack against Aspern. Archduke Charles, believing the decisive moment had arrived, personally led the assault. The fighting in the village was savage. There was hand-to-hand fighting in the cemetery, in the streets, in the houses. The village was taken and retaken six times. Finally, Molitor's four regiments were forced from the village. At this point, Legrand's division counterattacked, and Molitor's battle-weary men joined them. Part of the village was recaptured by the French. Reinforcements were arriving, and the main bridge over the Danube had been repaired. General Carra Saint Cyr's infantry division crossed over to the bridgehead at 6:00 P.M., followed an hour later by elements of Nansouty's and Saint-Sulpice's heavy cavalry divisions. Carra Saint Cyr went to support the fighting at Aspern; however, the entire village could not be retaken. Fighting there raged into the night, until it finally petered out with the French holding the eastern half of Aspern and the Austrians holding the western half. The new cavalry reinforcements joined with Bessières to try one last cavalry charge in the center, but this charge was thrown back.

On the Austrian left, Rosenberg finally got into position to attack Essling at 6:00 P.M. Charles had clearly weighted his right, making the main emphasis at Aspern, and Rosenberg's corps had only 21,000 men and fifty guns.[16] Boudet held Essling with 5,553 men and six guns.[17] As at Aspern, the Austrians first committed their artillery for a massive preliminary bombardment, and the barrage effectively ruined half the town.[18] Rosenberg launched three assaults against Essling, all of which were repulsed thanks to the strength of the buildings and the courage and leadership of the French. By nightfall, the fighting had stopped on the battlefield save for the firing still going on in Aspern.

The night's lull in the fighting offered Napoleon an opportunity that he failed to recognize. At the start of the day, the French had had 23,000 troops in the bridgehead and by late afternoon the numbers had risen to 31,500.[19] Charles had over 90,000 facing the bridgehead. Napoleon could cut his losses by evacuating the bridgehead under the cover of night, which would have been the wisest thing to do. What the Emperor should have realized was that his position was untenable. The French had fought a far stronger force to a standstill and had the advantage of the villages of Aspern and Essling to serve as bulwarks for the defense. However, the French could not remain in the bridgehead, for to stay there would doom them. Since they could not stay there and they were not going to retreat, they would have to attack. To attack, the French would have to emerge from the relative protection of the bridgehead into the open, presenting

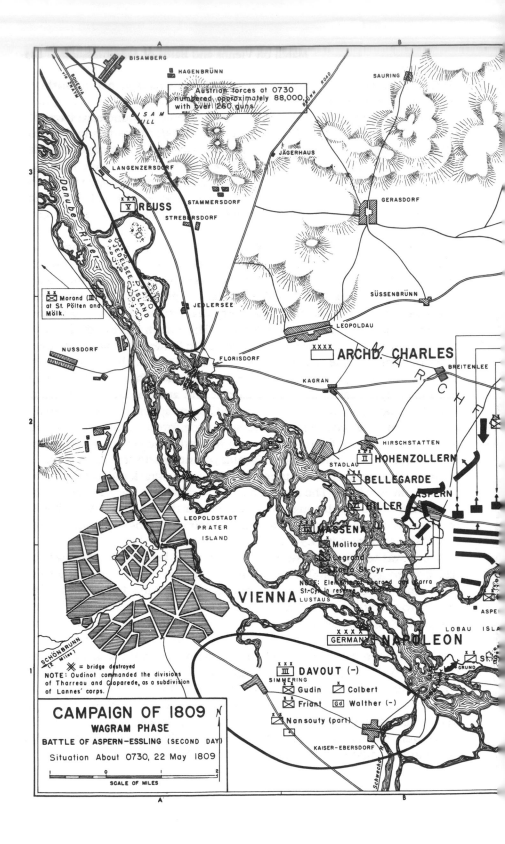

BISAMBERG

HAGENBRÜNN

Austrian forces at 0730
numbered approximately 88,000
with over 260 guns

SAURING

BISAM HILL

JÄGERHAUS

LANGENZERSDORF

GERASDORF

REUSS

STAMMERSDORF

STREBERSDORF

SÜSSENBRÜNN

Danube River

Morand (III)
at St. Pölten and
Mölk.

JEDLERSEE

LEOPOLDAU

NUSSDORF

FLORISDORF

ARCHD. CHARLES

BREITENLEE

KAGRAN

HIRSCHSTATTEN

HOHENZOLLERN

STADLAU

BELLEGARDE

ASPERN

LEOPOLDSTADT
PRATER
ISLAND

MASSENA

Molitor
Legrand
Carra St-Cyr
NOTE: Elements of Legrand and Carra
St-Cyr in reserve behind

VIENNA

LUSTAUS

ASPE

LOBAU ISLA

GERMAN NAPOLEON

GRUND

SCHÖNBRUNN
(2 Miles)

✗ = bridge destroyed

NOTE: Oudinot commanded the divisions
of Tharreau and Claparede, as a subdivision
of Lannes' corps.

DAVOUT (-)

SIMMERING

Gudin   Colbert

Friant   Gd Walther (-)

Nansouty (part)

KAISER-EBERSDORF

CAMPAIGN OF 1809
WAGRAM PHASE
BATTLE OF ASPERN-ESSLING (SECOND DAY)
Situation About 0730, 22 May 1809

0   1   2
SCALE OF MILES

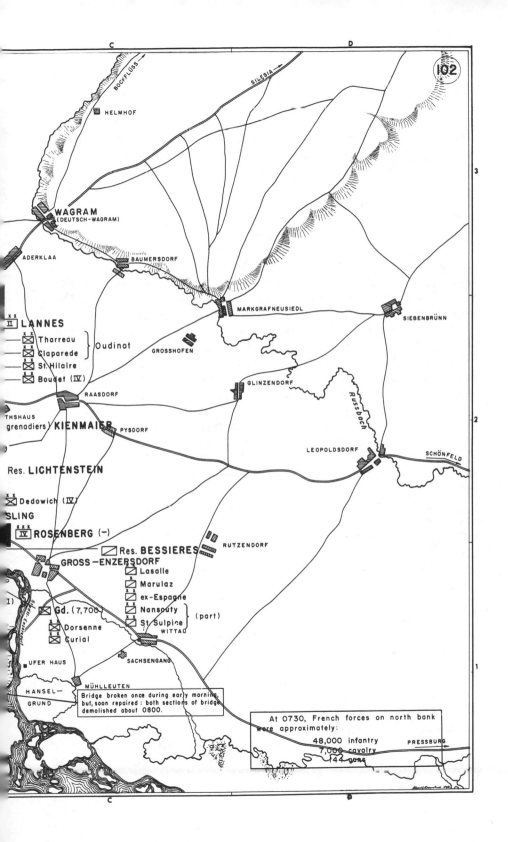

102

HELMHOF

WAGRAM
(DEUTSCH-WAGRAM)

ADERKLAA

BAUMERSDORF

MARKGRAFNEUSIEDL

SIEBENBRÜNN

GROSSHOFEN

II LANNES

— Tharreau ⎫
— Claparede ⎬ Oudinot
— St.Hilaire ⎭
— Boudet (IV)

GLINZENDORF

Russbach

RAASDORF

THSHAUS
grenadiers **KIENMAIER**

PYSDORF

LEOPOLDSDORF

SCHÖNFELD

Res. **LICHTENSTEIN**

Dedowich (IV)

SLING

IV ROSENBERG (-)

RUTZENDORF

Res. **BESSIERES**

GROSS-ENZERSDORF

Lasalle
Marulaz
ex-Espagne
Nansouty ⎫ (part)
St Sulpice ⎭

Gd. (7,700)

WITTAU

Dorsenne
Curial

SACHSENGANG

UFER HAUS

MÜHLLEUTEN

HANSEL—
GRUND

Bridge broken once during early morning,
but, soon repaired : both sections of bridge
demolished about 0800.

At 0730, French forces on north bank
were approximately:

48,000 infantry
7,000 cavalry
144 guns

PRESSBURG

of the village. Napoleon was forced to commit four battalions of the Imperial Guard to restore Essling and the French right.

The front of the bridgehead was but two and a half miles in total length. As the Austrians attacked, they were compelled to narrow their own front so the ratio in actual infantry combat between Austrians and French would be one to one. The battle at Aspern-Essling had become one of attrition, and Charles would have to use up a lot more of his own men to capture the bridgehead. At this point, Charles decided to spare his troops and destroy the French with his artillery. By 1:30 P.M., the entire French battleline was ringed with Austrian guns. The bombardment was all the French could stand.

The main bridge had been repaired by midday, but a floating burning mill sent downstream by the Austrians broke it again. Napoleon realized that he would have to retreat to Lobau, but to start a retreat now, in daylight, would be suicidal; they would have to hold until nightfall.

The Austrian cannonade continued throughout the afternoon. At 4:00, Marshal Lannes was struck in the legs by a cannonball, and carried from the field, he died from infection on May 31. He was among the boldest and most intelligent of Napoleon's commanders as well as one of his few real friends. Napoleon wept openly at his death.

The firing died down at nightfall, and the French retreat to Lobau Island began at 11:00. Massena, commanding the rear guard, crossed the pontoon bridge at 3:30 A.M. on May 23. The pontoon bridge was drawn back from the north bank to Lobau, and the last of the French infantry in the bridgehead got into boats and rowed ashore to Lobau. At 4:00 A.M. the Battle of Aspern-Essling was over.

The casualties for the Austrians in this battle were not light. In all, 4,286 were killed, 16,314 wounded, 837 taken prisoner, 1,903 missing. In addition, 2,048 horses were lost as well as one standard and six guns.[21] The French losses are more difficult to estimate. Generally, they range from a total of 20,000 to 30,000,[22] and the actual figure was probably closer to the latter.

Napoleon was stunned by the defeat and in an indecisive daze for the next thirty-six hours. He had been beaten on the battlefield for the first time in his life. The loss of so many good men, as well as some of his closest friends and best commanders, was a hard blow, made all the worse because it was inflicted by an army that he believed deserved nothing but contempt. Something had certainly gone wrong, but what?

The French official version blamed the defeat at Aspern-Essling on the breaking of the bridge, not the Austrian army. This is a *canard*. Some have argued that the French were defeated because Napoleon failed to build a protective screen for his bridges and French reconnaissance failed to locate the enemy army. But reconnaissance in war is rarely perfect. What actually happened at Aspern-Essling was that the Austrian army defeated the French.

The reason for the Austrian success was that the reforms instituted prior to the war were bearing fruit. The tactical performance of the Austrian infantry was good, and every French attack was eventually stopped. Of particular note was the ability of the Austrian units to form squares in the face of French cavalry charges. The seesaw battles for Aspern and Essling show that the Austrians were well-matched against the French. On the smaller battlefield of Aspern-Essling, the Austrians were able to effectively coordinate the movements of the different corps as well as use the three combat arms. The volume of artillery firepower on so small a front was something new on Napoleonic battlefields, but the comparative ease and liberal use of the guns indicated the effectiveness of the corps system and improvements within the Austrian corps, division, and brigade staffs. This was no longer the army of Austerlitz.

Yet the French were not destroyed. Is it fair to criticize Charles for not delivering the coup de grace (as Meade was later criticized for a similar failure at Gettysburg)? The Austrians had clearly tried to wipe out the French bridgehead, and the French position would have collapsed if either of the villages had been effectively taken. The small length of the front, combined with the physical layout and structure of the villages, favored the defenders. The Austrians did not have enough room to deploy their numerically superior forces and were compelled to attack frontally, engaging the French one for one. This was an attritional battle that would have led to the destruction of the French bridgehead given enough time. It was doubtful that a final assault late in the afternoon of May 22 could have carried the villages; having sustained a 23 percent casualty rate in twenty-four hours, the Austrian army was approaching exhaustion. Firepower and attrition, the characteristics of modern nineteenth-century warfare, were emerging.

When Napoleon recovered from his stupor, he realized he was facing an army that deserved respect. The volume of Austrian artillery fire impressed him, and he realized that the inequality in numbers of guns between his army and the Austrian had to be corrected. More men would be needed, too. What Napoleon did not realize, and would never grasp, was that as opposing armies modernized, the likelihood of winning a decisive battle diminished.

Charles may have been almost as stunned by his victory as his opponent. No one had ever beaten Napoleon before. Charles considered an attack on Lobau Island for the night of May 23 but abandoned the idea, believing the river too high to facilitate a crossing.[23] For Charles, this was just as well; the Austrian commander viewed the events of Aspern-Essling as a vindication of the strategy he adopted after Eckmühl. Rather than build on the victory, Charles preferred to stand pat, keep the army concentrated behind the Danube, and hope that the strain of maintaining the French army in the heart of the Danube valley would be too great for Napoleon, compelling him to withdraw.

In late May, the strategic situation presented an interesting picture. The two major armies seemed stalemated on the Danube near Vienna. Eugene de Beauharnais had breached the southern defenses of the Austrian Empire and was advancing for a junction with Napoleon (see Chapter 9). The British were preparing an amphibious operation in northern Europe to support the Austrians. The war against the French in Spain continued, and the Austrians could hope that their victory at Aspern-Essling would induce Prussia to join with them. The revolt in the Tyrol still burned, and there were a few minor revolts against the French in northern Germany.

What was happening in the east? Archduke Ferdinand commanded the Austrian VII Corps in Galicia. On April 17, Ferdinand invaded the Duchy of Warsaw and defeated a smaller force of Poles and Saxons under Prince Poniatowski at Raszyn. The Poles fell back to Warsaw, which they evacuated on May 23 for the right bank of the Vistula. Warsaw was occupied by the Austrians. Ferdinand continued north but was checked at Thorn and Plock. On May 5, Russia declared war on Austria in accordance with the Erfurt Treaty, but with wars of his own against Sweden and the Ottoman Empire, Tsar Alexander could send but a single expeditionary force to cooperate with Poniatowski. The Russians were commanded by Prince Gallitzin. The Poles and the Russians did not get along very well. Poniatowski believed, no doubt correctly, that the Russians were playing a double game. Meanwhile, Polish troops infiltrated into the rear of the Austrian corps and compelled Ferdinand to retreat. The Russians and the Poles invaded Austrian Galicia and jointly took the city of Cracow, but the erstwhile allies almost came to blows over control of the city. In short, the Polish theater was a sideshow that had little, if any, bearing on the events in Italy or on the Danube. But events in Italy would have an impact on the war along the Danube.

CHAPTER NINE

Junction

The Austrians needed to prevent the convergence of the Armies of Germany and Italy. Archduke John had a daunting task: to defend all of the Austrian frontier south of the Danube with limited means. His army had been badly beaten, and many of his best troops had been lost in Italy. He had Chasteler's and Jellachich's divisions in the Tyrol and Schmidt's brigade on the upper Piave. There were few trained reserves to be had in Carinthia or Carniola. The feudal levies in Croatia and Hungary had to be called up if there was any hope of securing the southern provinces of the empire.

There were many routes into the Carnic Alps by which men on foot could pass. However, armies consisted not just of infantry but of artillery, cavalry, and transports as well. To bring an army with all of its support into Carinthia meant traveling over the few primary roads capable of bearing the traffic. The Austrians had forts guarding the primary roads at Malborghetto and Predil and fortifications covering the road near Tarvis. If the French wanted to get to Villach and Klagenfurt, they would have to pass that way. If John's army had not been sent to Italy, he could have used the forts as a screen behind which he would have powerful mobile reserves that could fall upon the French columns moving up the narrow river valleys—but this reserve force did not exist. Of the 19,000 troops that returned to Tarvis with John, 5,000, the remnants of Ignatius Giulay's IX Corps, were sent back to Carinthia,[1] which was protected only by the brigades of Zach, Kalnassy, and Stoichewich. More troops were needed there to secure it and serve as a nucleus to raise the Croatian Feudal Ban. This left John with Albert Giulay's VIII Corps consisting of 13,000 demoralized regulars and some *Landwehr*. John expected Eugene to bring 30,000–40,000 troops against him. He believed that with so small and ragged a force, an effective frontier defense was impossible, so he would leave only a covering force on the frontier to delay the French for as long as possible. John would take his army deeper into the interior to gain reinforcements as he fell back and to cover the mobilizations in Hungary and Croatia. In addition, John ordered the withdrawal from the Tyrol of all Austrian regular troops. These were to join him in Austria, bringing 10,000–15,000 badly needed trained troops.

As Prince Eugene contemplated a mountain campaign, he knew that

CAMPAIGN OF 1809

WAGRAM PHASE

Situation Late 4 July 1809

SCALE OF MILES

dispersion, not concentration, was the key. His initial objectives were the Austrian bases of Klagenfurt, Villach, Tarvis, and Laibach. He deployed his forces along as broad a front as possible to stretch the Austrian forces to the breaking point and to outflank any strong points that would be encountered. Once through the mountain barrier, the army could reconcentrate. This approach owed its intellectual foundations both to Napoleonic practice and to Bourcet's concepts laid down in the *Principes de la guerre des montagnes*.

The Viceroy ordered Rusca, who was already on the upper Piave, to continue north into the Carnic Alps and then turn eastward to threaten Tarvis by way of the Gallitz River valley. Eugene himself would take a powerful column on the main road from Osoppo to Tarvis by way of Malborghetto. Totaling 25,000 men, this force included the Light Division, the VI and XII Corps, the Royal Guard, the artillery reserve, and the cavalry divisions of Sahuc and Grouchy.[2] Another column, consisting of Serras' division, was to take the route up the Isonzo toward Predil and continue from there to Tarvis. Collectively, the three columns totaled 36,000. Macdonald had a different assignment.

The Viceroy, like John, had to consider Dalmatia and Croatia. Marmont was still under Eugene's command, and little word had been heard from his detached force for some time. Marmont could be in need of help. Moreover, Eugene needed to secure his operational right as he advanced into Austria, and Italy had to be secure while the bulk of his army moved north. Consequently, he ordered Macdonald to take his corps east to Carniola and capture the Austrian base at Laibach. From there, Macdonald was to either support Marmont or turn north to seize Marburg and Graz. This scheme adhered in part to the envelopment envisioned by Napoleon in his offensive plan for Italy. It offered the other advantage of flanking the Austrian defenses between the Tagliamento and the Isonzo. Eugene's army would advance on an operational front over a hundred miles in length.

To achieve his mission, Macdonald had the divisions of Lamarque and Broussier and was reinforced with Pully's cavalry division and three *voltigeur* battalions from Dessaix's command. In all, Macdonald had 14,000 troops.[3]

Rusca moved his division up the Piave pursuant to his orders. The upper Piave was defended by Schmidt's brigade, which had been detached from the main body since the Austrian army had evacuated the Brenta. Schmidt had been methodically destroying bridges and cutting down trees as he withdrew up the river valley to delay Rusca's advance. Eventually, the road became impassable to wheeled vehicles, and Rusca realized he would have to leave his artillery and transport behind if he wanted to advance further. This he was not prepared to do, so he was compelled to reverse his march and follow the route of Eugene's column.

Eugene was also having difficulties. His march was halted by the broken bridges over the Fella between Venzone and Ponteba. The infantry could cross the river in rafts, but those vessels were too small and light to take artillery and cavalry. Eugene's engineers estimated that it would take 1,200 workmen at least five days to rebuild the bridges. Not wanting to suffer such a delay, Eugene sent his wagons, cavalry, and artillery, except for a few light guns, back the way they came and ordered them to follow the route of Serras' division. The rest of Eugene's column would continue to Malborghetto.

Albert Giulay had 6,000 men around Tarvis; however, they were men of little training and low morale. Beyond conducting a static defense, not much could be done with these troops, and any sort of offensive maneuver was out of the question. Malborghetto had a garrison of 650 men and ten guns while Predil was held by 250 Croatian infantry and eight guns. Advance units of the French column reached the two forts on May 15.[4]

Malborghetto was imposing enough to give the Viceroy pause. He was not going to attack carelessly; the fort would be surrounded before being attacked. Malborghetto commanded the main road, but there were mountain tracks over which infantry could bypass the fort. Baraguey d'Hilliers, with the divisions of Dessaix and Fontanelli, was sent over these tracks to block the main road from any attempt at relief from Tarvis. Grenier's corps would take the fort. Grenier sent General Michel Pacthod, who had taken command of Abbé's division, over the mountain tracks to attack the fort from the far side while Durutte's division attacked from the near side. The coordinated assault was launched at 9:30 A.M. on May 17. It was all over in thirty minutes; 300 Austrians were killed and 350 captured at a cost of 80 Frenchmen.[5] The spoils of victory included thirteen guns and enough food to feed a division for a week. Eugene pressed on to Tarvis.

Giulay had fought several skirmishes with Baraguey d'Hilliers' corps on May 16 and then evacuated Tarvis, taking a defensive position east of the town. A line of redoubts and redans erected east of the town along the banks of the Schlitza and Galitz streams was to include twenty-four guns, but only ten had been emplaced so far.

Eugene's attack began at midday. Baraguey d'Hilliers' corps attacked the flank of Giulay's position while Grenier's corps engaged the front. Fontanelli's division stormed a redoubt that anchored the Austrian left flank. The Italian division then began to roll up the Austrian line while Grenier's troops came into action. His position turned, Giulay ordered a retreat that disintegrated into a rout. The Austrians lost 3,000 men and most of their artillery; Eugene's forces lost but 300.[6] Eugene wanted to pursue, but without cavalry this was impossible. Predil had to be taken so Eugene could get his guns and horses up the main road.

Serras had bombarded Predil throughout May 17 with little effect, and Eugene ordered him to storm the fort the following day. The Viceroy sent three battalions south from Tarvis to attack the fort from the north while Serras assaulted from the south. The Croatians put up a desperate fight, but the entire garrison was wiped out while the French lost 150 men.[7]

Archduke John had been at Villach with 11,000 troops since May 15, and he had decided to withdraw to Graz even before Eugene attacked Tarvis. Albert Giulay was ordered to bring his forces to Villach, but these orders arrived too late to be carried out; Giulay was already heavily engaged and forced to retreat eastward down the valley of the Save. John evacuated Villach and marched to Graz, which was reached on May 24.

Eugene captured Villach and Klagenfurt on May 19 and 20, respectively. Supplies captured at those locations would support the army for the next stages of the advance. However, it was time to call a halt. The roads behind were not totally free from obstacles, and straggling had become a problem. Troops, horses, and transports clogged the roads all the way back to Osoppo and Caporetto, and it would take several days for Eugene's artillery to reach Villach. If Eugene wanted to have an effective military force for the next stage of the advance, he would have to wait for the tail of his army to catch up with its head. It was during this lull that Dessaix's Light Division was dissolved.

Dessaix's command had been created to rapidly cross rivers to catch the enemy, and that mission had been accomplished. Besides, three of the *voltigeur* battalions already had been sent with Macdonald, reducing its elite status. To compensate for the loss, Dessaix had been reinforced with some line units, but now the divisional commanders wanted their battalions back. The campaign had been a strain on the troops, and losses from all causes were mounting—a typical division was averaging 5,000–6,000 men. Eugene agreed to return the units to fill up the ranks of the other divisions, and Dessaix was given a brigade in Durutte's division.

The next stage of the advance to Bruck would take the Army of Italy into the Austrian province of Styria. There were no enemy units blocking the way, but there were enemy forces on both flanks. The enemy units to Eugene's left were the divisions of Chasteler and Jellachich, Schmidt's brigade, and the Tyrolean partisans. The threat from the Tyrol had diminished. Lefebvre's intervention into the area had drawn Chasteler away from Italy. Chasteler, along with the partisans, was attacked and severely mauled by Lefebvre on May 13 at Wörgel. Lefebvre went on to temporarily repress the Tyrolean insurrection, and Chasteler received orders from John to evacuate the Tyrol and join him.

Jellachich had been part of the flank guard of Hiller's VI Corps and was

cut off when Charles' army was driven north of the Danube in April. Jellachich had been unengaged. Originally he moved toward the Tyrol but retreated eastward, as did Charles' army. Eventually, Jellachich, too, was ordered to effect a junction with Archduke John.

Schmidt's brigade was falling apart. Demoralization spread in his ranks
as he retreated up the Piave. French intelligence reports indicated that
only half of Schmidt's 3,000 men still carried arms. The defeat of Chasteler and the disintegration of Schmidt's command indicated to the Viceroy
that the forces on his left were not particularly dangerous but needed to
be watched. Rusca's division had joined Eugene's column, so the Viceroy
gave it the mission of holding the Drave Valley between Villach and Klagenfurt and providing security from any threat from the Tyrol.

The forces on Eugene's right were more dangerous. Archduke John was
moving on Marburg where he would be rejoined by Albert Giulay and
other reinforcements, raising his strength to about 15,000. To the southeast was Ignatius Giulay's IX Corps, which was being reinforced by Croatian irregulars. Eugene organized a task force under General Grouchy to
secure his right flank.

Grouchy, with Pacthod's infantry division and half the light cavalry under Sahuc, was to march his corps down the valley of the Drave to Marburg. From there he would protect Eugene's flank from any interference
from John and cover the advance of Macdonald from Laibach.

The attempt to cover Macdonald's advance came in response to new orders from Napoleon. Eugene was ordered to hurry his march and bring as
many men as possible to join the Army of Germany. Consequently, Macdonald was not to wait at Laibach for Marmont; instead, he was to move
north as soon as possible. His route of march would take him to Marburg,
where he would join Grouchy, and together both of their forces would
move to Bruck by way of Graz and the valley of the Mur. The combined
forces of Grouchy and Macdonald would total 20,000. Grouchy's task
force reached Marburg, which was evacuated by John, on May 24, and a
junction was made with advance units of Macdonald's corps the following day.

The Viceroy, with Grenier's corps, the First Dragoon Division, Baraguey d'Hilliers' corps, and the Royal Guard, set out for Bruck via Saint Michael and Leoben on May 23. His route would bring on an encounter battle with Jellachich's division, which wanted to link up with John in
Hungary. Jellachich was trying to get to Graz down the Mur River valley
via Saint Michael and Bruck. This seemed like the safest route since the
Army of Italy was already in the valley of the Drave, and Napoleon was
sending units southward from Vienna to probe toward Bruck. Moreover,
the road that runs through Saint Michael was the only one in the area suit-
able for heavy artillery. With a little luck, Jellachich could move through

Saint Michael and Bruck before it was blocked by French troops; unfortunately, Jellachich's luck ran out.

Jellachich collided with the Army of Italy at Saint Michael on May 25 and took up a defensive position south of the town; he had 8,400 troops. Eugene brought up Grenier's corps, totaling 12,000. The French engaged Jellachich on his front while turning both his flanks. A final combined arms assault shattered the Austrian line. In all, 800 Austrians were killed, 1,200 wounded, and 4,200 captured at the cost of 600 French casualties.[8] An immediate pursuit by the French cavalry toward Leoben brought in an additional 600 prisoners. Only 1,600 of Jellachich's troops ever managed to reach John's army; the division had ceased to exist as a fighting force.

Although a minor battle in and of itself, the destruction of Jellachich's division at Saint Michael had an important impact on the course of Eugene's later operations in Hungary. Austrian resources were being stretched to the limit, and most of Jellachich's command had consisted of frontline regulars as opposed to the poorly trained and equipped *Landwehr* and Hungarian levies that were reinforcing John's army. What John needed most were regular troops, and the 8,000 regulars that Jellachich might have brought could have made a difference at the Battle of Raab in June and in the course of operations on the Danube. The day after the Battle of Saint Michael, Eugene, with Serras' division in tow, reached Bruck and direct contact was made with Napoleon's army. The junction between the armies of Germany and Italy was accomplished.

How had the forces on Eugene's right been faring? Marmont had been acting independently since the start of the war. Taken by surprise by the Austrian offensive into Dalmatia on April 10, he withdrew to Zara according to plan. After uniting his forces, he returned to the offensive and severely mauled Stoichewich's brigade at Mount Kitta on May 13. Two further battles were fought at Gradschatz on May 17 and Gospich on May 21, leading to the almost complete destruction of Stoichewich's command. Moving north, Marmont took Fiume on May 28 and reached Laibach on June 3.

After being detached from the main body, Macdonald reached Udine on May 12, capturing large amounts of Austrian supplies as well as many Austrian sick and wounded. Palmanova was also relieved. From there, Macdonald headed for the Isonzo and Carniola. The retreating brigades of Zach and Kalnassy were too weak to stop the French, and meeting light resistance, Macdonald forced the Isonzo near Goritzia on May 14. From there, he pushed on to the Austrian frontier fort at Prawald, where the Austrian garrison was so demoralized it surrendered to the French on May 20. The Austrian main base at Laibach was also taken with ease on May 23; Ignatius Giulay and his reinforcements arrived too late to save it. The capture of Laibach and Prawald placed 7,000 muskets, seventy-one guns, and great quantities of food and ammunition into Macdonald's hands.[9]

It was at this point that Macdonald received Eugene's orders to bring his corps immediately to Bruck via Marburg and Graz. He set out north, leaving a garrison to hold Laibach and await Marmont. Macdonald joined Grouchy at Marburg, and together they reached Graz on May 29.

Not wishing to face the combined forces of Grouchy and Macdonald, John evacuated Graz and marched east to the Raab River. John left a garrison in Graz, but this force was too small to hold the entire town, and the Austrian commander negotiated a convention with Macdonald: the French would take Graz, but the Austrians would hold the town's citadel. Both sides were to refrain from any military actions inside the town; any attacks made against the citadel would have to be mounted from outside the city limits. Macdonald kept his corps at Graz to besiege the citadel while the rest of Grouchy's command marched to Bruck.

After Macdonald relieved Palmanova, the French garrison there marched south to take the port city of Trieste, where they captured 22,000 muskets that were being delivered from the British to arm the Croatian and Hungarian feudal levies.

Strategically, the offensives by the Napoleonic Armies of Germany and Italy were simultaneous advances aimed at the heart of Austria. Two deep penetrations had been driven into the Austrian Empire to converge at Vienna, one running from Ratisbon, the other from Tarvis. The advance of the Army of Italy effectively cut Austrian communications via the Drave and Mur rivers, and key Austrian bases were captured. The speed of the French advance managed to disrupt Austrian positions in the Tyrol as illustrated by the fate of Jellachich's division. The course of the war was characterized by nonlinear operations. By driving salients into enemy territory, both Napoleon and Eugene had to look to their flanks and rear as well as to their front to maintain their position and momentum. Up to this point, the French armies had been operating on exterior lines. The junction of the two armies ensured that they could support each other on interior lines while the Austrian armies of Charles and John were compelled to remain distant, operating on exterior lines.

The Army of Italy had been engaged in constant operations since leaving the Adige, and four weeks of campaigning had taken its toll. The average strength of Eugene's infantry units had dropped from 6,000–7,000 per division to 5,000–6,000. Many horses had died during the month, immobilizing transport and artillery, and the French were compelled to requisition horses in Styria to make up the losses. Eugene did not expect to have his army assembled and replenished until June 1.

After reaching Bruck, the next several days were spent in organizing the rear areas and dealing with logistical and manpower problems. The

French made a practice of utilizing the resources of occupied territories to support their war machine, and Styria, Carinthia, and Carniola were no exception. To help pay for the cost of the war, financial contributions were exacted from the local populace. For example, a tax of 50 million francs was levied on the city of Trieste.

The spoils of war included 140 guns captured by Eugene's army. These would be critically needed to redress the inferiority of French artillery firepower compared to that of the Austrians, and the captured pieces were added to the divisional and regimental artillery reserves.

The last days of May were spent drawing supplies and men. Klagenfurt was established as a logistical base for the Army of Italy. To fill up Eugene's depleted divisions, the rear areas of the army were combed to find more troops. Serras' division was moved to Neustadt to be closer to Vienna should Napoleon suddenly need reinforcements. For the remainder of the army at Bruck, it was a time of rest.

After setting up the administration of the conquered territories, Eugene left Bruck on May 29 to meet with Napoleon at Vienna. The time had come to plan for the final defeat of the Austrians, and the Army of Italy would play a critical role in the forthcoming operations.

CHAPTER TEN

The Wagram Campaign:
The First Phase

The junction of the Armies of Germany and Italy ended one part of the war and marked the beginning of another. The Italian and Alpine campaigns had been won by the French, but Napoleon's campaign on the Danube had been stalled due to his defeat at Aspern-Essling. A new campaign to win the war was about to begin.

Modern military campaigns are conducted in phases, and each major operation or series of operations constitutes a phase. Such was the case in this war and particularly in the events of June and July 1809. Napoleon's strategic objective was the destruction of Archduke Charles' army. How was it to be achieved? The French army still had to cross the Danube safely, the communications of Napoleon's army had to be secured, and once across the Danube, the Austrians had to be defeated in battle. To do this, Napoleon had to accomplish certain things. The first was to ensure that when the French met the Austrians again, they would have tactical equality, if not superiority. Napoleon had been shaken by the intense firepower the Austrians had brought to bear at Aspern-Essling. Clearly, the French had to match or surpass the power of the Austrian guns, which meant the French artillery had to be increased. Writing to Eugene, Napoleon said he believed that he had faced "close to 400 guns" at Aspern-Essling.[1] On May 24, as soon as his wits had returned to him, he set about increasing the firepower of his army. Napoleon ordered captured Austrian artillery to be sent to the corps so that every infantry regiment would have two guns attached to it.[2] This would give each infantry brigade from four to six guns to effectively match the Austrian infantry brigades in firepower.

Even these additions were not enough, so more guns were brought from everywhere in Napoleon's empire.[3] Eugene was ordered, "to bring as much artillery as possible"[4] and to requisition extra guns from the captured Austrian bases at Klagenfurt. As a result of these efforts, by July 3 Napoleon's forces had 683 guns.[5]

To defeat Archduke Charles, Napoleon would have to take as large an army as possible across the Danube; he would not be outnumbered again. The arrival of Eugene and the Army of Italy brought a welcome reinforcement, and Napoleon would use Bernadotte's IX Corps and possibly some of Lefebvre's VII Corps for his next operation. To move north of the

Danube, the rear of Napoleon's army and that of Eugene's had to be secure, and it was also necessary to prevent Charles from being reinforced. Herein lies the sequencing of operations for this campaign: the Austrian forces south of the Danube had to be dealt with prior to any offensive north of it.

Enemy forces still operated south of the Danube and threatened French communications. Chasteler's division was in the Tyrol as were the Tyrolean partisans; they were being contained by Lefebvre's VII Corps. Archduke John's army in Hungary, however, posed a far greater threat.

Since leaving Graz and moving toward Körmend behind the Raab River, John's army had slowly been growing in strength. With the survivors from Italy serving as a nucleus, John had been reinforced with both regular Austrian units and *Landwehr*, increasing his troop strength to about 20,000.[6] Eighty miles north of Körmend was the fortress city of Raab (Gyor) at the confluence of the Raab, Rabnitz, and Little Danube rivers. It was there that about 10,000[7] largely untrained levies of the Hungarian *Insurrectio* had been gathered under the leadership of Archduke Joseph, the Prince Palatine of Hungary. About 100 miles to the south of John's position was Ignatius Giulay's IX Corps at Agram (Zagreb) on the Save River, with an additional 8,000 troops.[8]

The three groups at Raab, Körmend, and Agram were under John's overall command and totaled approximately 40,000. By themselves, these three separate bodies did not constitute a major threat, but if they were united into a single mass and flung against Napoleon's operational right flank and rear, they could jeopardize his offensive against Charles. What caused Napoleon further concern was that by holding the Danube River crossings at Pressburg, Raab, and Komorn, John could bring all of his forces north to reinforce Charles and nullify the numerical superiority created by Eugene's reinforcement of Napoleon's army. Even if John's forces moved north of the Danube, they could still pose a threat by launching sorties south of the Danube by way of Pressburg, Raab, or Komorn. Consequently, Napoleon wanted to neutralize the Austrian forces in Hungary and seize control of the river crossings as a prelude to his own attack against Charles.

These operations were entrusted to Davout and Eugene, and Davout's crack III Corps had the easier task: to capture the fortified bridgehead opposite Pressburg. The more difficult operation, that of neutralizing John's army and securing the crossings at Raab and Komorn, was given to Eugene, which proves that Napoleon had confidence in his stepson's abilities as a military commander. In spite of his early defeat at Sacile, Eugene had produced impressive victories in Italy and Austria, and it would have been just as easy for Napoleon to have sent Davout to the Raab and Eugene to Pressburg. By assigning the more difficult Raab operation to

Eugene, the Emperor showed he considered the Viceroy to be the equal of the marshals.

After leaving his army at Bruck, Eugene spent two days with Napoleon at Vienna discussing the upcoming operation. Napoleon stressed several factors: the importance of Eugene being able to countermarch quickly to Vienna for the decisive battle with Charles, the capture of the town of Raab, and the neutralization of John's army, either by destroying it in combat or by driving that army far enough away to prevent it from interfering in the battle against Charles. What was impressed most upon Eugene at this conference was the fact that his operation was to be totally subject to Napoleon's larger strategic goals and that he should always be ready to conform his own operations to Napoleon's higher goal, the destruction of Charles' army. Consequently, the overall direction of the Hungarian operation was under Napoleon's control. This was facilitated by the proximity of the Hungarian front, which was but twenty-four to forty-eight hours riding time from Napoleon's headquarters. Thus, the Emperor would be able to transmit orders to Eugene much more quickly than when the Viceroy was in Italy.

Eugene left Vienna on June 2 and rejoined his army at Neustadt, where it had been moved in order to be closer to Napoleon's army. At Neustadt were the forces of Grouchy, Grenier, Baraguey d'Hilliers, and the Royal Guard—26,000 troops in all. Macdonald's corps was still at Graz, and Marmont was moving on Laibach. Rusca's division was still holding Klagenfurt.

In preparation for his advance on Hungary, Eugene was ordered to move to Oedenburg, which he reached on June 5. Napoleon considered Eugene's army at Oedenburg too weak for its mission and so placed General of Division Count Jacques Lauriston's mixed division, which had been screening Eugene's army to the northwest, under the Viceroy's command. Lauriston's mixed force consisted of a brigade of Badenese infantry and a light cavalry brigade under General of Brigade Pierre Davis, the Baron Colbert-Chabanais, and the addition of Lauriston's forces brought Eugene's effective strength to a total of 30,000.[9]

Napoleon wanted to include Macdonald's corps in Eugene's operation, but the Emperor also wanted to capture the citadel of Graz. Therefore, Macdonald was to leave Broussier's division to besiege the fortress while Macdonald joined Eugene with the 7,000 troops of Lamarque's and Pully's divisions.

Neither Napoleon nor Eugene knew the exact location of John's army. The Austrians were last reported to be in the vicinity of Körmend, but it was not known if they were still there. Consequently, Napoleon was uncertain as to the direction in which Eugene should march and decided to keep him near Oedenburg until John's army could be found. Reconnaissance patrols were sent by Colbert and Pully to scout the banks of the Raab River.

On June 5, Colbert's troopers captured an Austrian soldier and sent him to Viceregal Headquarters. The prisoner reported that only a rear guard was at Körmend and that John's main body was moving toward Raab to join with Archduke Joseph. Since Eugene was under Napoleon's direction, he could not act on this information without consulting Napoleon; he could, however, make suggestions. The Viceroy proposed that if this information could be confirmed, he should march directly eastward on Raab with the troops of Grouchy and Grenier and intercept John en route while the rest of his army would protect his and Napoleon's flank and rear.

Eugene's proposal is interesting in that it reveals his understanding of Napoleon's system of war. Eugene knew that John's army was the main target and that it should be destroyed as quickly as possible. In a letter to Napoleon, he pointed out that if he marched southeast on Körmend, he would only be striking thin air if John was already moving north.[10] But by moving immediately eastward to Raab, Eugene hoped to interpose his army between John and Joseph, secure a central position, and use interior lines to defeat both in detail. Furthermore, Raab was closer to Vienna than Körmend, and Eugene could countermarch to Vienna more quickly from Raab should Napoleon need him. A significant flaw in Eugene's plan was that he proposed taking only 17,000 men with him for the operation.

Napoleon considered Eugene's plan too reckless, vetoed it, and sent him his own thoughts in a letter dated June 6:

> First of all, you must march assembled and concentrated. I don't believe that Serras, Durutte, and Grouchy's five cavalry regiments are sufficient. Baraguey d'Hilliers' corps and the Royal Guard should be with you so that you will have 30,000 men at hand marching united in such a way as to be assembled on the field in three hours time.[11]

Napoleon evidently considered Eugene's proposal a product of Eugene's recent Austrian campaign, so he sought to point out the difference between Eugene's last campaign and the present one and to outline his own plans for operations in Hungary.

> In the plains of Hungary, one must maneuver differently than in the gorges of Carinthia and Styria. In Carinthia and Styria, if one overtakes the enemy at a point of interception as at Saint Michael for example, the enemy column is destroyed. But in Hungary, on the contrary, as soon as one occupies one point, the enemy will quickly move to another. Suppose that the enemy is moving on Raab, and

that you arrive there before him, the enemy, learning of it en route, will change direction and march on Pesth [Budapest].

What would the enemy do in this situation? Should he abandon Styria, Carinthia, Giulay's corps and all the south of Hungary, uncover Pesth before the movements of Macdonald and Marmont in order to cross to the left bank of the Danube? Or should he, to the contrary serve as a nucleus to unite all of the Hungarian Insurrection, rally the troops who have fled before Marmont, interrupt your communications with Laibach, and cover Pesth, which is after all, the capital of Hungary. In this last case, it would seem possible that the enemy would maneuver on Körmend behind the Raab River, interrupt communications from Graz to Laibach, and still hold himself ready to cover Pesth. Then your movement on Raab would take you away from him, and would give the enemy the idea (for he unlike us, is in familiar surroundings and is better informed) of attacking and overthrowing Macdonald. I think therefore that a movement on to Stein am Anger and Körmend, or from Guns to Savar would be wise, if you haven't any other information than what I have at this moment.

As for me, it does not appear to be yet proven whether the enemy is retiring on Komorn or on Raab. I think he will remain in observation, and will act according to the maneuvers we make against him, always keeping open his retreat on Pesth, and that if he retires on Raab, it would be better to outflank him on his left, than on his right, for by this means you would pass the river [the Raab] toward Sarvar, and would throw him into the Danube; for at Komorn as well as at Raab, he requires at least three days to cross the Danube; and finally, by this maneuver, you protect General Macdonald and General Marmont, and you can unite with them.[12]

Two elements stand out from these instructions. First, Napoleon envisioned one of his famous *manoeuvres sur les derrières* against John's left via Körmend. Second, any such movement would be limited by Napoleon's concern about John turning his own operational right. This concern had kept Eugene's army close to Napoleon's during the first few days of June, was no doubt discussed at Vienna, and appears in Napoleon's correspondence as early as June 4.[13] Napoleon's desire to keep Eugene nearby to cover his flank and rear would limit the depth of Eugene's turning maneuver against John's left. Hence, Eugene might strike John's flank but would not be able to drive beyond it into his rear, which was the real object of a *manoeuvre sur les derrières*. By so limiting Eugene's operational vista, Napoleon would help ensure the escape of a large portion of John's army.

It is evident by this exchange of ideas that Napoleon did not keep his methods of war a secret from his subordinates. Eugene was not treated as an automaton, ordered to march his army from one point to another without getting a glimpse of the master plan. Rather, the entire concept of operations was revealed from supreme commander to army commander. Different operational approaches were discussed, and it is interesting that the Viceroy proposed his own plan of operations. Notice that enemy options were considered in Napoleon's instructions, as they were in previous operational plans sent to Eugene and Berthier prior to the start of the war. In the June 6 letter, Napoleon also informed Eugene that he would be reinforced with the light cavalry division under Montbrun and that Eugene was to take his army from Oedenburg to Guns in preparation for a drive on Körmend via Sarvar.

One point that should be noted is that both Napoleon and Eugene expected Marmont to be in supporting distance within a week. His army corps had reached Laibach by June 3, and it was not unreasonable to hope that Marmont could be at Graz or even Körmend by June 10. However, Marmont was delayed by the threatening movements of Giulay's IX Corps and Chasteler's division. Consequently, Marmont did not actually join the main French armies until the end of June and so had no real part in Eugene's operation.

On June 6, information began to pour into both Imperial and Viceregal Headquarters concerning the location of the major Austrian units south of the Danube. It was confirmed that Chasteler was moving down the valley of the Drave, threatening Villach and Klagenfurt. Rusca was at Klagenfurt and, it was hoped, could deal with Chasteler. Eugene received news that his previous information concerning John's movement was wrong. The Archduke's main body was at Körmend, with other units at Saint Gothard and Furstenfeld. Giulay's corps, joined by the remnants of the Austrian forces from Dalmatia, was moving north from the Drave River. All evidence pointed to a general concentration of these three groups at Körmend.[14]

Eugene estimated that if Chasteler could bypass Rusca, he could arrive at Körmend between June 10 and June 12 and that Giulay could be there on June 15 or 16. Relying on the operational direction imposed by Napoleon's June 6 letter, Eugene planned to attack John with superior forces before Chasteler and Giulay could arrive. A concentration of over 40,000 men was ordered to take place in the area between Sarvar and Körmend by June 9. Eugene marched out from Guns for Sarvar with Lauriston's division as an advance guard, followed by the forces of Grouchy,[15] Grenier, and Baraguey d'Hilliers. Macdonald was ordered to leave one division (Broussier's) at Graz to continue the siege of the citadel and to march with the rest of his corps (Lamarque and Pully) toward Körmend. Marmont was to hold himself in readiness to deal with either Chasteler or Giulay depending on circumstances.

However, the hoped-for battle at Körmend did not take place. Realizing that he was outnumbered, John evacuated his positions upon learning of Eugene's advance and moved northeast toward Papa; from there he could continue toward Raab and Joseph's Hungarian levies.

Eugene reached Sarvar on June 9, but Macdonald was nowhere to be found because the latter only received his orders to march on June 9 and so could not reach Körmend until late the following day. The Viceroy had also lost contact with John, so Eugene organized a force to find the enemy. Grouchy was given a large task force consisting of the cavalry divisions of Guerin and Montbrun, the latter being reinforced by Colbert, and the infantry of Serras and Lauriston. Grouchy crossed the Raab at Sarvar and probed to the east below Papa.

Meanwhile, Chasteler, whose presence in the Tyrol had influenced Eugene's operations in Italy, would also influence the current campaign. Chasteler emerged from the Tyrol with 4,000–5,000 troops[16] and attacked Rusca at Klagenfurt on June 9. Rusca held the town but was unable to prevent Chasteler from bypassing Klagenfurt and continuing eastward down the Drave to join with Giulay.

The passage of Chasteler's division had for a time disrupted communications with Italy, and Napoleon was worried that the combined forces of Chasteler and Giulay would return to threaten the rear of his own army. To prevent such an occurrence, Marmont was ordered to reinforce Broussier at Graz. Such a force would have been strong enough to deal with Chasteler and Giulay, but Napoleon was so sensitive about any threat to his lines of communication that he ordered Eugene not to heavily engage Macdonald's corps lest that general's troops be needed to face Chasteler and Giulay.[17]

Napoleon's order put Eugene in a quandary as to the use of Macdonald's corps. He knew that his own operation was part of a larger plan to win the war, and he knew from his past conversations and correspondence that Napoleon was much concerned about any disruption of his lines of communication. If Eugene chose to commit deeply all of his forces to destroy John, he might win a battle at the cost of upsetting the Emperor's other plans for the campaign. Eugene decided to leave Macdonald in the rear, ready to move at the Emperor's call. Macdonald's troops would support Eugene in case of a reverse, but Eugene felt constrained from using them in an offensive manner. This stricture concerning Macdonald's corps would ensure that Eugene's victory over John on June 14 would be incomplete.

Grouchy found John's army, which was moving north toward Papa, on June 11. His prey located, Eugene set his whole army in motion. Speed was of the essence since Eugene wanted to destroy John before he could combine with the troops of the Hungarian *Insurrectio* at Raab. Grenier and Baraguey d'Hilliers marched northeast from Sarvar to join Grouchy before Papa on June 12. Macdonald, who had now reached Körmend, was ordered to cross the Raab and its tributary the Marczal and turn north to follow the army toward Papa. Eugene did not expect Macdonald to join him until June 13, but with the rest of his army closing on Papa, he thought he had enough troops to beat John.

Early on June 12, John left a rear guard at Papa and headed north toward Raab before Eugene's army arrived. Grouchy's cavalry arrived in front of Papa early in the afternoon, and under the direct orders of the Viceroy, the massed French squadrons smashed through the screening Austrian cavalry and stormed into the town. Most of the rear guard fell or was captured by the charging French horsemen. However, John and his main body had managed to get away.

Eugene was furious that his adversary had escaped. Intelligence reports conflicted as to whether John was heading for Raab or Komorn. Raab was closer to Vienna, and Eugene, influenced by Napoleon's instructions not to move too deeply into Hungary, decided to move on Raab rather than Komorn.

The town of Raab is approximately twenty-five miles north of Papa and is located at the confluence of the Raab River, which runs roughly from south to north; the Little Danube River, which flows from largely west to east; and the Rabnitz River, which runs parallel to the Little Danube. To the east of the town, between the Little Danube's southern bank and the Raab's eastern bank, was an entrenched camp where Archduke Joseph's troops of the Hungarian Insurrection were gathered. Raab itself was a major road junction for the routes running along the banks of the Raab and Little Danube rivers.

Three to four miles south of the city is a long ridge that runs largely perpendicular to the main highway on the Raab's eastern bank. The ridge, known as the Csanak Heights, begins two-thirds of a mile from the river bank and extends two miles in an easterly direction. The village of Csanak lay on the northeastern slope of the ridge.

Two and a half miles north and east of the Csanak Heights is a large plateau called the Szabadhegy Heights, whose northeastern face was a mile and a half from the town of Raab, and the village of Szabadhegy was located on and around the western side of the plateau. In front of the center of the southern face of the plateau, located on a slight rise of ground, was the Kismegyer farmhouse, surrounded by a high, thick stone wall. Three hundred yards due east of the farmhouse was a steep mound topped by a stone chapel, and a mile and a half further east was the Pancza marsh,

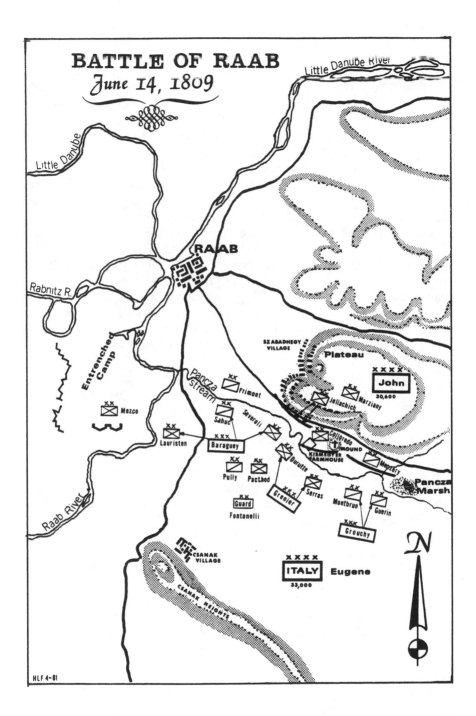

BATTLE OF RAAB
June 14, 1809

Little Danube River

Little Danube River

Little Danube

Rabnitz R.

RAAB

Entrenched Camp

Mezco

Pancza Stream

Frimont

Sahuc

Severoli

Lauriston

Baraguey

Pully

Pacthod

Durutte

Guard

Fontanelli

Grenier

Serras

Montbrun

Guerin

Grouchy

SZABADHEGY VILLAGE

Plateau

John
30,600

Marziany

Jeliachich

Colloredo MOUND

KISMEGYER FARMHOUSE

Mecsery

Pancza Marsh

CSANAK VILLAGE

CSANAK HEIGHTS

Raab River

ITALY
33,000
Eugene

N

HLF 4-81

from which the Pancza brook flowed in front of the mound, farmhouse, and plateau to empty into the Raab River.

John arrived at Raab on June 13 with thirty-five battalions and twenty-eight squadrons.[18] In an entrenched camp next to Raab, Archduke Joseph had assembled thirteen battalions and thirty-eight squadrons.[19] The combined forces now under John's overall command totaled 35,528 and 30 guns.[20] Only half of the troops were regulars, the rest of the army consisting of *Landwehr* and Hungarian levies, who were imperfectly trained and armed; this was particularly true of the Hungarians.

Upon his arrival at Raab, John received orders from Archduke Charles commanding him to send 8,000 of his troops to Pressburg, which was being threatened by Marshal Davout.[21] John was to base the rest of his army at Raab and Komorn to harass Eugene to keep the Viceroy as far away from Napoleon as possible; after this John was to move to the north bank of the Danube and join Charles above Vienna.[22] John was not too happy with these orders, especially with the prospect of losing his status as an independent commander by uniting with Charles.[23]

While John was considering these orders, Eugene's vanguard was seen approaching from the southwest. John had the option of crossing the Little Danube at Raab or Komorn, but he chose instead to stay on the southern bank of the river and fight.

John had fought two major engagements against Eugene, at Sacile and the Piave, and had won the first. Now, for the first time since the battle on the Piave, John could meet Eugene with a force approaching numerical parity. However, because of the large number of "green" *Landwehr* and Hungarian levies in John's army, he did suffer from a qualitative inferiority compared to Eugene's veteran troops. John had received large amounts of new artillery and hoped that this, in conjunction with a proper use of terrain, would compensate for his qualitative inferiority. Both the Csanak and the Szabadhegy positions could provide the natural basis for a strong defense, upon which Eugene's army could be broken in an assault against either of the two heights. If so, a victory over the French would enable John to regain contact with Giulay and Chasteler and, he hoped, allow him to continue to operate south of the Danube largely independent of Charles' control.

Evidently, John decided to base his defense along the Csanak Heights. Four Austrian battalions were already around the Heights, and John wanted to occupy that place in force before Eugene, but it was Eugene who got there first. The Viceroy, riding with Grouchy's cavalry corps, swept the Austrian infantry from the area and seized the ridge. When a strong enemy force, consisting of six cavalry regiments, an infantry brigade, and fifteen guns, was seen advancing from the north, Eugene ordered Grouchy to charge the oncoming Austrians to delay their advance until Eugene's own approaching infantry and artillery could occupy and secure the Csanak Heights.

Grouchy launched a series of charges against the advancing Austrians and drove off the enemy cavalry. The infantry, which stayed in squares supported by their guns, did manage to advance slowly. But the French cavalry had done its job, and the infantry of Lauriston and Durutte, plus Sorbier's artillery, managed to secure the heights before the Austrians could arrive.

John wanted Eugene to attack him in a strong position, not the other way around. The Archduke did not think his army, with its many raw troops, was capable of mounting a major assault to capture the Csanak Heights, so John recalled his attack force and decided to make his stand based on the Szabadhegy plateau.

The Szabadhegy position was in many ways superior to the Csanak Heights for a defensive battle. The Szabadhegy plateau commanded all of the southern and western approaches to Raab, the circular shape of the plateau ensured that an attack from any direction would have to move up-hill, and the plateau was wide enough to allow for the easy deployment of large masses of troops. The village of Szabadhegy, located on the western half of the plateau, would favor the defenders as they could use the houses for cover. The southern approach to the plateau was protected by the Kismegyer farmhouse and the chapel mound, and both of those places could serve as strong points to break up any enemy attack (just as Hougomont and La Haye-Sainte would later do at Waterloo).

John hoped that Eugene would attack these strong points and get bogged down there, after which the French could be thrown back by his infantry massed on the plateau. John would use his cavalry to cover the areas between the plateau and the Raab River to his right, and between the plateau and the Pancza marsh to his left.

It was now late in the day, and the rest of the afternoon and evening of June 13 was spent in preparation for the upcoming battle, with John seeking to occupy his chosen position and amalgamate his Austrian and Hungarian units. In John's army, the old corps structure had disappeared. The Austrian regular and *Landwehr* units had been combined into divisions of two to three brigades each, and there were three infantry divisions commanded by Jellachich, Colloredo, and Marziany. During the day, eight of the Hungarian battalions were brought over from the entrenched camp to reinforce the divisions of Jellachich and Colloredo on the plateau.

In all, John had forty-three battalions massed on the plateau and sixty-six squadrons protecting his flanks. Jellachich's division would hold the western half of the plateau, which included the village of Szabadhegy as well as a nearby bridge over the Pancza brook; Colloredo would hold the eastern half as well as the farmhouse and mound. Marziany's division would be held in reserve behind the other two. On John's left, between the mound and the Pancza marsh, there would be forty squadrons, most of them Hungarian, under Archduke Joseph and Generalmajor Meczery (Meczery was the actual commander of the left wing; Joseph was merely a figurehead). Frimont would hold the ground between Szabadhegy and the Raab with twenty-six squadrons consisting of both Hungarian and Austrian troops. On the opposite bank of the Raab was the entrenched camp, held by five battalions and six squadrons, all Hungarian, under Generalmajor Mezco. It would be Mezco's job, once Eugene advanced against the plateau, to threaten Eugene's left and rear.

Eugene was not certain whether John would attack him the following day and so spent the rest of June 13 positioning his army on and around the Csanak Heights with his left resting on the Raab River. Eugene's entire army was now present on the field, save for Macdonald's corps, which had just reached Papa. Orders were sent to Macdonald to march to the Raab area as soon as possible.[24] There was some trouble in locating Macdonald, since the general had retired for the night in a house outside the town. By the time Macdonald was found, it was already after midnight, and it took some time to get the corps on the road. Pully's division, being mounted, managed to join Eugene before dawn, bringing Eugene's total troop strength to 27,982 infantry, 10,299 cavalry, and 42 guns.[25] Macdonald, accompanying Lamarque's infantry, did not arrive until 4:00 P.M., when the battle was all but over.

As the morning of June 14 wore on, it became apparent to Eugene that John was not going to attack. Eugene knew that John's army posed a threat to Napoleon's plans and if he could destroy that army by battle, he could ease the situation for the Emperor and pave the way for the trans-Danubian offensive against Charles. Eugene therefore decided to attack and destroy John.

Eugene believed himself to be outnumbered (he thought John had 40,000 troops),[26] but he knew that one-third to one-half of John's army consisted of raw troops and untried levies and that the battle power of John's army was nowhere near what it had been in Italy. Eugene believed that his own veteran army could beat John's even without Macdonald's corps. The Viceroy was under orders not to engage Macdonald's troops heavily, and as afternoon approached, Eugene feared that John might withdraw if he delayed any further. So Eugene decided to begin the battle without waiting for the arrival of Lamarque's division.

Surveying the Austrian positions, Eugene decided to turn John's left and throw him into the Danube, just as Napoleon hoped he would do. Eugene would use the corps of Grenier and Baraguey d'Hilliers as the *masse primaire* to frontally assault the Austrians on the plateau and force the Austrians to commit their reserve. The *masse de manoeuvre* would consist of

Grouchy's two cavalry divisions under Guerin and Montbrun, which would attack Meczery's cavalry between the march and the mound and then pivot inward toward the eastern face of the plateau. Once John's forces were stretched to support his left, the *masse de rupture*, consisting of Pacthod's infantry division, Fontanelli's Royal Guard, and Pully's dragoons, would attack, breaking the Austrian line. This approach bears remarkable similarity to Napoleonic practice and Eugene's approach at the Piave.

The Army of Italy attacked at noon.[27] The French infantry received a warm welcome from the Austrian guns, and fighting was heavy in the village of Szabadhegy, at the farmhouse, and on the mound. Grouchy's cavalry negotiated the Pancza brook and was promptly attacked by Meczery's cavalry. The rival horsemen charged and countercharged each other, but the fire from the French horse artillery batteries enabled Grouchy to gain the upper hand.[28] By combining fire and shock, Grouchy drove the Austrian cavalry back, unmasking the left flank of Colloredo's infantry division. Guerin was sent to follow Meczery's cavalry and keep them at bay, and Montbrun was sent against Colloredo's left flank. This threat forced Colloredo to evacuate the mound and bend back his line to face Montbrun. John also sent some units from Marziany's division to the eastern face of the plateau to ward off Montbrun. Serras' infantry division occupied the mound when the Austrians withdrew, and Serras then moved to attack the farmhouse. The divisions of Severoli and Durutte drove Jellachich's troops from Szabadhegy and advanced to the crest of the plateau.

John had evidently hoped that he could maintain his position while the French dashed themselves to pieces against his defenses. Anything more elaborate was beyond the capacity of his troops. But the advance of Severoli and Durutte to the plateau threatened to crack his line, and he needed veteran troops to reestablish his position. Jellachich was reinforced with three battalions from the reserve and three more from Colloredo, and the veterans led a bayonet charge that drove the French off the crest and back toward the Pancza brook.

Reserve units from Severoli's division contained the Austrian counterattack. Eugene ordered up Pacthod's fresh division to attack Jellachich, after which he personally rallied the troops of Severoli and Durutte and led them back into the fight. The three divisions of Pacthod, Severoli, and Durutte hurled back Jellachich's forces and stormed the crest of the plateau. By this time, Serras had taken the farmhouse and was driving Colloredo's weak and exhausted troops up the slope of the plateau.

John still had the ten fresh battalions of Marziany's division, but they represented his last infantry reserve. The French were now on the plateau in force, John's left flank was threatened, and the troops of Jellachich and Colloredo were largely spent. Eugene still had the Royal Guard in reserve, and another infantry division—Lamarque's—could be seen approaching in the distance. If Marziany were committed to a counterattack and repulsed, there would be nothing left to cover the retreat, and John's entire army would be lost. This battle of attrition was going in favor of the French. Deciding that discretion was the better part of valor, John used Marziany's division and as much cavalry as he could muster to cover the retreat of his army to Komorn.

Marziany managed to hold back the French while the troops of Jellachich and Colloredo streamed past in retreat. Some of Frimont's cavalry withdrew into Raab while the rest fell back eastward to support Marziany. Sahuc and Pully were sent to follow Frimont while Lauriston closed in on the suburbs of Raab. Frimont's cavalry was eventually driven off, and Sahuc's horsemen closed in on Marziany's infantry. Several infantry squares were broken, and Marziany himself was captured.

However, the coordination of the French pursuit slackened. Grouchy did not have command over Sahuc and Pully, and many divisional and regimental commanders did not keep a tight enough rein over their troops. The result was that by the time the French cavalry got itself sorted out, night was falling and John's army, along with the bulk of the rear guard, had managed to escape. John's army continued to retreat eastward throughout the night. Upon reaching Komorn the following morning, John crossed to the north bank of the Danube, leaving a strong garrison holding the bridge at Komorn.

While the French army followed John's retreating forces, the Hungarian units on the left bank of the Raab abandoned their camp. Some went into Raab to garrison the town, and the rest moved south down the left bank of the river toward Sarvar to raid Eugene's communications. This effort proved to be merely a nuisance, since Eugene's communications were being rerouted along the southern banks of the Rabnitz and Danube rivers to Vienna.

The Battle of Raab cost the Austrians 6,000 casualties, three guns, and two colors.[29] French casualties totaled 2,500, but General Severoli was among the wounded.[30] Generals Baraguey d'Hilliers, Grenier, Durutte, Serras, Dessaix, and General of Brigade Bonfanti all had horses shot out from under them. Eugene had won a tactical victory leading to an incomplete success. John had suffered damage—he retreated to the north bank of the Danube and was separated from Giulay and Chasteler—but his army was not destroyed. Could it have been?

The Austrians had a strong defensive position, but even though the numerical difference between the two forces at Raab was not great, the qualitative disparity was. A major part of John's army consisted of barely trained or totally untrained men, whereas Eugene had all veterans. It was for this reason that the French could drive the Austrians away. However, there was enough structure in John's army for it to stay in existence. The Austrian rear guard helped the main body escape, and there was no Austrian collapse.

Grouchy's corps, although used as a *masse de manoeuvre*, conducted a tactical turning movement; Grouchy did not execute a *manoeuvre sur les derrières*. Raab was a frontal battle, not a battle of envelopment. Indeed, the Viceroy could not conduct an envelopment with the forces he had with him at Raab, because only Macdonald's corps at Papa was in a position to execute a deep maneuver that could envelop the Austrian army. Constrained by Napoleon's orders not to heavily commit Macdonald in case he might be needed to countermarch back to Graz to support Broussier or Marmont, Eugene had to keep Macdonald close to the Raab River. If sent deep into Hungary in order to envelop John, Macdonald would be unavailable if Napoleon suddenly needed more troops. Eugene clearly understood that his own operation was part of a larger plan of campaign. Such a clear understanding of one's role in a campaign serves as a model for the nineteenth and even the twentieth century, for such understanding is still rare.

In the context of the overall campaign, the operation from June 5 through June 14 was only partially successful in achieving Napoleon's goals. Eugene did succeed in pushing John north of the Danube and separating him from Chasteler and Giulay, thus reducing the danger to Napoleon's operational right and paving the way for Napoleon's crossing of the Danube. But the French were unable to achieve their other goal of destroying John's army.

In spite of Eugene's limited success, Napoleon was more at ease about the Austrians south of the Danube, and Eugene was given full control of all of his forces, including Macdonald's corps. Eugene was ordered to capture Raab and destroy the fortified bridge at Komorn to prevent John from recrossing to the southern bank of the Danube. The Viceroy also was ordered to give the impression that he would soon march on Budapest in the hope that Austrian troops would move further east and away from the main target, Charles' army.[31]

General Lauriston, with his Badenese troops, was charged with the task of besieging and capturing the town of Raab. The siege of the city was carried out in an orderly manner, and the 3,500-man garrison surrendered to Lauriston on June 23 after a breach had been made in the walls.[32]

While Lauriston was besieging Raab, Eugene was busy observing Chasteler to the south (for this he had been given another cavalry division under Lassalle) and trying to destroy the bridge at Komorn. Chasteler was kept at bay, but the destruction of the bridge could not be accomplished. Because of the need to protect the siege operation from Chasteler, Eugene was reluctant to move on Komorn in force. Several attempts were made to break the bridge there by floating windmills and heavily laden barges down the river, but to no avail.

Napoleon considered sending Eugene's army to Komorn in force to destroy the bridge, but he abandoned the idea. By late June, Napoleon was preparing his great offensive against Charles and soon would want Eugene's army near Vienna; a major commitment of Eugene's army at Komorn could delay that offensive. Davout had also failed to secure Pressburg, and Napoleon did not want to spend any more time trying to take that bridgehead either. The Emperor believed that with Raab now in French hands, the town could be used to anchor his operational right and a small corps of observation could be kept there to contain any threat from Komorn or Pressburg.

What had been happening at Graz? Giulay, with 15,000 troops,[33] half of them irregulars, made a thrust against the city on June 25. Giulay drove back Broussier and resupplied the Austrian garrison in the fort, but Marmont finally moved into the area and, in concert with Broussier, inflicted 2,000 casualties on the Austrian IX Corps and drove it southeast to Gnas.

By June 28, the Austrian forces in Hungary, particularly the forces under John and Giulay, had been either mauled or kept at bay. The first phase of the campaign was completed, and Napoleon now felt secure enough about his flank and rear south of the Danube to launch his grand attack against Charles.

CHAPTER ELEVEN

The Wagram Campaign:
The Second Phase

While the first phase of the campaign moved forward, Napoleon was preparing for the second. The Emperor had to ensure a safe and secure crossing of the Danube. Six weeks were spent planning and preparing, and Lobau Island was turned into an advance base and fortress. After Aspern-Essling, Massena and his IV Corps remained on Lobau as a garrison. Fortifications were erected on the island, good roads were built to facilitate the movement of troops and supplies, and Lobau was defended by a great battery of 129 heavy guns.[1] The link between Lobau and the southern bank of the Danube was strengthened by building two strong bridges to link Kaiser-Ebersdorf with Schneidergrund Island and three more to link Schneidergrund and Lobau. A stockade was pile-driven into the Danube upstream from Lobau to catch any floating rams sent downstream by the Austrians to damage the bridges, and a flotilla of gunboats manned by Marines of the Guard patrolled the river as well. In addition, pontoons and other building materials were stored, ready to be floated into place and erected to allow the army to cross from Lobau to the north shore when the time came. Napoleon wanted a rapid crossing for his army. Sixteen separate bridges would be built by the time the crossing was completed.

On the Austrian side, Charles' victory at Aspern-Essling did not make him sanguine about the prospects of winning an offensive against the French. He even may have hoped that he would never again have to fight another big battle. Charles was in a position to draw on the resources of Upper Austria, Bohemia, and Moravia, and more reinforcements were brought in the form of Kollowrat's III Corps and 60,000 *Landwehr*, which were brigaded among the regular troops in Charles' army.[2] In addition, 200 more guns were added.[3] These and other reinforcements brought Charles' strength up to 142,000 troops and 414 guns by the end of June.[4] Charles could also call upon Archduke John and his army to join him, which would add an additional 13,000–20,000 men. What to do with this armament?

No attempt was made to significantly reorganize the army or to improve its tactics beyond what had already occurred. Charles did not attempt to encourage any initiative among his corps commanders other than stating that their main concern should not be the perfect alignment of their forces in the battle line.[5] The commanders were never taken into Charles' confidence, and centralized command and control remained the order of the day. Charles lacked the nerve to try an offensive operation. What was worse, unlike Napoleon he had no clear objective in mind other than preserving the army and maintaining the current situation. Perhaps this is why he was aloof from his subordinates: he had nothing to say.

Charles ordered the Aspern–Gross-Enzersdorf line fortified, which would surely halt the French should they try to copy their previous crossing, but beyond that he did nothing. He could not make up his mind if he should stay close to the river, to meet head-on any further attempt to cross it, or keep his troops further in the rear to engage the French after they had crossed. If he chose to wait, then what battle plan to adopt? Charles had difficulty in deciding and so from the beginning abandoned all of the initiative to Napoleon.

The positions taken by Charles' army were defensive. Count Klenau's Advance Guard and Hiller's VI Corps manned the fortifications from Aspern through Gross-Enzersdorf, and the rest of the army was deployed in two wings. One wing was behind the Russbach, defending the main route to Moravia; the other wing was on the Bisamberg Heights, protecting the main road to Bohemia. While Charles idled, Napoleon's preparations continued.

The Emperor wanted nothing left to chance concerning the crossing of the river, and he had no intention of crossing where the Austrians expected him to do so. He planned to deceive them by feigning another assault from the north shore of Lobau while the real crossing would be launched from the eastern shore of Lobau, depositing the French army below Gross-Enzersdorf and in a position to flank the Austrian fortifications. Napoleon and Massena made a personal reconnaissance of the proposed crossing sites by posing on the riverbank as common soldiers preparing to take a bath. The French and Austrian sentries had adopted a live-and-let-live approach in this area, so the two could examine the area at their leisure.[6] All was ready for the second phase of the campaign to begin.

On June 29, orders went out from Imperial Headquarters to concentrate the French forces for the decisive battle. Davout and Eugene were to bring their respective commands from Hungary by July 2,[7] and Eugene was ordered to leave Baraguey d'Hilliers, along with the Badenese troops and Severoli's division, to hold Raab and to observe Komorn and Pressburg. Broussier and Marmont were ordered to leave Graz and join the army at Ebersdorf; Rusca was to take over the duties of watching the Austrian-held citadel at Graz as well as holding Klagenfurt and Bruck and guarding communications with Italy. Bernadotte's IX Corps was brought in from Linz as was Wrede's Bavarian division, which was detached from the VII Corps. The remainder of Lefebvre's VII Corps would protect Napoleon's communications south of the Danube, and Vandamme and his VIII Corps would garrison Vienna. To put further pressure on the Austrians, Jerome and his X Corps were ordered to move through Saxony and threaten Bohemia.

The deception operation began on June 29 as well. As mentioned, there was a series of small islands in the middle of the Danube, and some of them were between Lobau and the north bank of the river. The French gave these islands the code names Massena, Bessières, Espagne, Pouzet, Lannes, and Alexandre. On June 29, Napoleon sent troops in assault boats from Lobau to reoccupy the Mühlau salient. The operation was covered by the fire of thirty-six guns, and a bridge connecting the salient to Lobau was rebuilt. On the first of July, Napoleon seized Bessières Island, which was just off the north shore of Lobau, and began to build a bridge there. Meanwhile, the French batteries on the north shore of Lobau began bombarding the Austrian fortifications between Aspern and Essling. On the evening of July 1, the II Corps (commanded by Oudinot since Lannes' death) marched to Lobau Island. Stores and materiel were brought over on July 2, and at 8:30 P.M. on July 3, the Imperial Guard moved to Lobau, followed by Bernadotte's IX Corps. This corps went to the northeast part of Lobau Island and pretended to prepare to cross the river at Muhlau; this was part of the deception plan. By July 4, Davout's III Corps was at Ebersdorf; so was the Army of Italy, but its complement was not complete. Eugene had with him Grouchy's cavalry corps, consisting of Grouchy's own division plus that of Pully and Sahuc; Grenier's corps, with the divisions of Durutte and Serras; the Royal Guard, and Macdonald's corps, with only Lamarque's division—the infantry divisions of Pacthod and Broussier having been delayed. On July 4, more transports, stores, and artillery were taken over to Lobau.

On the evening of July 4, under the cover of darkness, Massena's and Oudinot's corps moved to the east side of Lobau. Davout's corps crossed over to Lobau as well and took up position between Massena and Oudinot. At 1:00 A.M. on July 5, Eugene's army crossed to Lobau, followed by Bessières' Cavalry Reserve Corps, and four hours later, Marmont's corps followed. It had been raining heavily on July 3 and 4, and the rain and the darkness served to mask the French movements.

The movements and feints had the desired effect. The Austrians were aroused and looking intently toward the north side of Lobau Island. Learning of French actions on June 29 and 30, Charles ordered his entire army to advance to the Aspern–Gross-Enzersdorf line. Beyond this, Charles and his two chief advisers, Grunne and Wimpfen, were uncertain as to what to do. Observing the great artillery battery on Lobau, the Austrians decided that fighting near the riverbank exposed to the fire of those

Sommerin
IV: 7,000
observing Linz)

Schistek
(V: 7,000 at Krems)

BISAMBERG

HAGENBRUNN

III KOLLOWRAT
(16,600)

BISAM
HILL

SAURIN

Prochaska (grenadiers)

JÄGERHAUS

RES LICHTENSTEIN
(18,600)

V REUSS
(7,600)
LANGENZERSDORF

VI KLENAU
(13,800)

This corps was largely dispersed
along the river, observing crossing points

HAMMERSDORF

STREBERSDORF

GERASDORF

Aspre (grenadiers)

IX BERN
(18,4

LEFEBVRE
(14,900) at Linz.

JEDLERSEE

Lasalle
SÜSSENBRÜNN

Legrand

NUSSDORF

FLORISDORF

Marulaz

Leopoldau
Carra St-Cyr

IV MASSENA
(29,000)

Molitor
BREITENLEE

KAGRAN

Boudet

HIRSCHSTATTEN

STADLAU

ASPERN

LEOPOLDSTADT
PRATER

WÜRT.

VANDAMME
(10,000)

Vandamme was charged
with the occupation of
Vienna and the security
of the river line in the
vicinity.

VIENNA LUSTAUS

ASPER
Reynier
(4,500)
LOBAU ISLA
Began crossing to
bank about 1800.

Reynier had 7 infantry battalions
(Berthier's Neufchatel contingent
and six battalions drawn from the
II, IV, and IX Corps), plus considerable artillery.

Line of piles to
protect bridges.

SCHÖNBRUNN
(2 Miles)

SIMMERING

Pile bridges

LOB-GRUND

NOTE: ⌐⌐⌐⌐ = General line of Austrian intrenchments

CAMPAIGN OF 1809
WAGRAM PHASE
BATTLE OF WAGRAM (FIRST DAY)

Situation About 1400, 5 July 1809, and the
Crossing of the Danube by Napoleon's Leading
Elements on the Night of 4-5 July

N

0   1   2

SCALE OF MILES

KAISER-EBERSDORF

XII MARMONT

Wrede (VII)

Italy

Pacthod

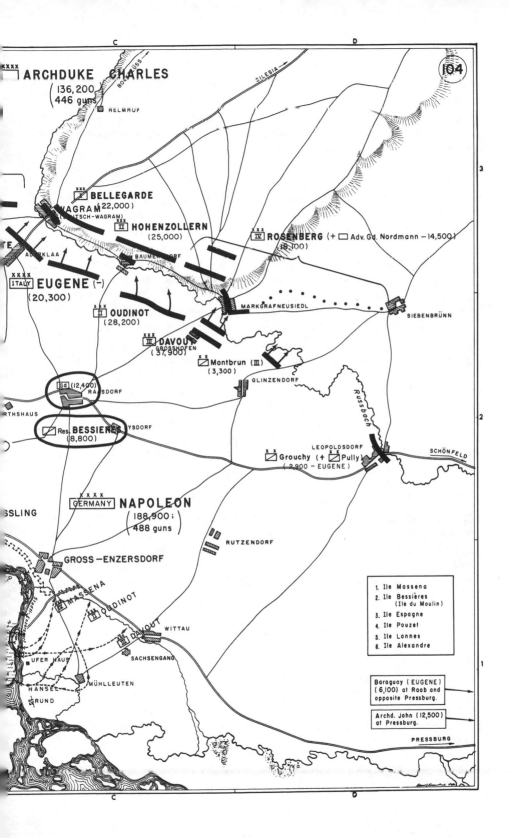

104

XXXX ARCHDUKE CHARLES
(136,200
446 guns)

III HELMHOF

BODEN RUSS
SILESIA

XXX
I BELLEGARDE
(22,000)
WAGRAM
(DEUTSCH-WAGRAM)

XXX
II HOHENZOLLERN
(25,000)

XXX
IV ROSENBERG (+ □ Adv. Gd. Nordmann — 14,500)
(18,100)

XXXX
ITALY EUGENE (-)
(20,300)

ADERKLAA
BAUMERSDORF
MARKGRAFNEUSIEDL

SIEBENBRÜNN

XXX
II OUDINOT
(28,200)

XXX
III DAVOUT
GROSSHOFEN
(37,900)

XX
Montbrun (III)
(3,300)

GLINZENDORF

Russbach

Gd (12,400) RAASDORF

WRTHSHAUS

Res BESSIÈRES
(8,800)
...YSDORF

LEOPOLDSDORF

SCHÖNFELD

XX
Grouchy (+ XX Pully)
(2,900 — EUGENE)

XXXX
GERMANY NAPOLEON
(188,900;
488 guns)

SSLING

RUTZENDORF

GROSS—ENZERSDORF

MASSENA

II OUDINOT

III DAVOUT
WITTAU

UFER HAUS

SACHSENGANG

HANSEL
GRUND

MÜHLLEUTEN

1. Ile Massena
2. Ile Bessières
    (Ile du Moulin)
3. Ile Espagne
4. Ile Pouzet
5. Ile Lannes
6. Ile Alexandre

Baraguay (EUGENE)
(6,100) at Raab and
opposite Pressburg.

Archd. John (12,500)
at Pressburg.

PRESSBURG

guns would be suicidal. It was better to fall back and attack the French after they had crossed. Therefore, on July 3, the Austrian army withdrew
save for the Advance Guard and VI Corps, which were to hold the Aspern–Gross-Enzersdorf line for as long as possible. Klenau now commanded the VI Corps, for Hiller had resigned, claiming ill health, and
Feldmarschall-Leutnant Armand von Nordmann had been given command of the Advance Guard. The rest of the army fell back to take positions behind the Russbach stream or in the Bisamberg Heights. Believing
that action was imminent, Charles ordered fortifications to be dug along
the Russbach, but this was too late to do any good. He also sent a dispatch
to Archduke John telling him to bring his army to the Marchfeld.

The simultaneous actions of July 4 and 5 constituted a marvel of detailed staff planning. Napoleon had conceived of the movement of the
army to Lobau, the deception plan, and the eventual crossing from the
east face of Lobau, but it was executed by the brilliant staff headed by
Marshal Berthier. The construction of sixteen bridges and the movement
of over 180,000 troops with their guns and supplies across the river and
into an immediate assault was a project on a scale that had never before
been attempted. Yet it was done. From the Mühlau salient the French had
captured some prisoners who gave them a general idea about the location
of the Austrian army. By the evening of July 4, everything was ready for
the assault.

Napoleon ordered a night assault to ensure surprise. There was a tremendous thunderstorm during the night of July 4, and the heavy rain
helped mask the French movements. At 9:00 P.M. the 129 French guns on
Lobau opened fire on the Austrian defenses north of Lobau. The fire of
the guns combined with the thunder of the storm produced an incredible
din. The hurricane of shot kept the Austrians huddled in their defenses
and seemed to presage an assault from the north side of Lobau. In the
meantime, the crossing from the east face of Lobau began.

Napoleon had envisioned a broad-front attack in which Massena,
Oudinot, and Davout would launch their forces in succession.[8] Between
9:00 and 10:00 P.M., infantry from Oudinot's corps got into assault boats—
escorted by gunboats of the river flotilla—crossed to the north bank of
the river at Hansel-Grund. The average distance across the river was 125–
179 yards, and there was a strong current. The fire from the gunboats
drove off the few Austrian sentries on the north shore, an area that was
not strongly held. The Austrian command did not believe that a crossing
would be made from the east side of Lobau, because Napoleon would
have to form a front parallel to his line of communications. This was a
conservative estimate that discounted Napoleon's determination or his in-
genuity.

After the leading infantry secured the crossing points on the north bank, pontoon bridges were floated downstream and swung into place. At 10:00 P.M., Massena began a similar crossing. Napoleon was so excited that he rode over to watch the crossing in Massena's sector, where a bridge consisting of fourteen pontoons was being readied to be swung into position to span the 179-yard channel. The Emperor turned to one Captain Heckmann, an engineer officer in charge of the operation, and asked, "How long do you require for the swinging?" "A quarter of an hour, Sire." "I give you five minutes. Bertrand, your watch."[9] The bridge was in position in four minutes, and Napoleon was urging the infantry across even before it was fully in place.[10] The building of additional bridges continued throughout the night and into the following day.

With the bridgeheads secure, Oudinot advanced to take Muhlleuthen and Massena, Ufer Haus. By 5:00 A.M., Davout's corps had crossed, but there was a bit of a mix-up at this point. Davout had been located between Massena and Oudinot, but the Emperor had wanted the IV and III Corps to form, respectively, the left and right wings of the army. Troops from the III Corps began to get mixed up with the II Corps as the two units marched to their positions, but the staff officers managed to sort things out. At 9:00 A.M., Massena attacked the Austrian forces holding Gross-Enzersdorf from the south and east and cleared the town. By 10:00 A.M. on the morning of July 5, the first echelon of the army, consisting of the corps of Massena, Oudinot, and Davout, was in position facing north with Wittau on the right and Gross-Enzersdorf on the left. In all, 95,000 troops were in position to expand the bridgehead while the second echelon consisting of the Army of Italy, Bernadotte's IX Corps, Bessières' Cavalry Corps, and the Imperial Guard began to cross to the north bank.

The weather had cleared after the stormy night, and the crossing of the Danube and the deployment of the first echelon must have been an awe-inspiring sight. The tricolor battle flags marked each regiment in a sea of blue uniforms and glittering bayonets, as the finest army in the world advanced to the beat of hundreds of drums.

The first echelon rolled up the Austrian defenses from Gross-Enzersdorf to Aspern. By 2:30 P.M., Massena had driven Nordmann from Essling, and the Austrian commander had taken his forces north of the Russbach. Massena continued to Aspern, which was held by Klenau's VI Corps, and had taken the town by 4:30 in the afternoon, forcing Klenau to retreat to the Bisamberg Heights. As the French advanced north and east, their front expanded, and the gaps in the line were filled by the second-echelon forces as they came up.

Exclusive of the Lobau garrison, Napoleon committed to the north bank 180,490 troops and 488 guns,[11] although this total was not reached until the following day. In late afternoon on July 5, the French army approached the line of the Russbach.

As has been mentioned previously, the Marchfeld was a broad, flat area bordered by the Russbach stream and a series of hills. Within the Marchfeld were a number of villages, including Raasdorf, Breitenlee, Süssenbrünn, Gerasdorf, and Aderklaa. The villages of Leopoldsdorf and Baumersdorf straddled the Russbach, and those of Wagram and Markgrafneusiedl were immediately behind the stream on its northern bank. There was a ridge behind the Russbach that ran four miles in length between Markgrafneusiedl and Wagram, at which points the flanks of the ridge turned north. The top of the ridge was only ten feet above the bank of the stream. To the east of Wagram was a series of hills terminating at the Bisamberg Heights overlooking the Danube.

If one measures in a straight line from Leopoldsdorf to Bisamberg, the distance is fourteen miles. This is an enormous battle area and one that dwarfed all previous Napoleonic battlefields, encompassing twenty-five square miles. One finds it impossible to walk the Wagram battlefield as easily as one can walk Waterloo or Gettysburg. Unlike other Napoleonic battlefields, it was impossible for either Napoleon or Charles to effectively see all of the forces under their command. The forces moving toward the clash on the Marchfeld were also the largest ever seen in European history up to that point and would be surpassed in the nineteenth century only by the battles at Leipzig in 1813 and Königgrätz in 1866. Because of the immense size of the armies and the battle area, the influence of the respective army commanders waned while the daring and the initiative of the corps commanders was of greater significance than at Aspern-Essling.

Archduke Charles' army was deployed in a vast arc twelve miles in length from Markgrafneusiedl to Bisamberg. His left wing was on the Russbach Heights with Rosenberg's IV Corps and Nordmann's Advance Guard at Markgrafneusiedl, Hohenzollern's II Corps in and around Baumersdorf, and Bellegarde's I Corps holding Wagram and the immediate vicinity. This wing totaled about 79,000 troops.[12] To the east of Wagram was Lichtenstein's Reserve Corps of grenadiers and heavy cavalry deployed between Wagram and Gerasdorf. Further to the east, among the Bisamberg hills, were the corps of Klenau (VI), Kollowrat (III), and Reuss (V). Other than calling for John to march to him, Charles' battle plan was to await events, although a more mature plan would develop as the evening wore on.

Napoleon's battle orders were all verbal, so an accurate picture has to be pieced together. The orthodox view is that Napoleon realized Charles' army was overextended and that he should attempt to split that army in half by penetrating east of Wagram.[13] If this was true, the assaults conducted on July 5 made little sense. Given the size of the battle area, it was very unlikely that any French reconnaissance could pick up and effectively estimate the location and size of Charles' right wing (III, V, and VI Corps). From the deployment of the Army of Germany, as will be shown below, Napoleon sent the bulk of his forces against the Russbach position and would leave his left flank dangerously exposed. It seems quite probable that Napoleon was unsure of the location of all of Charles' forces and believed that most of his army was behind the Russbach. Napoleon was aware that John was a threat to his western flank and sent a column to the west to keep watch for him. Napoleon's battle plan may not have gone further than to attack the Russbach position simply to develop the situation; in any event, he ordered the attack at 6:00 P.M.

By that time, Napoleon's army was deployed as follows: Grouchy's cavalry corps had been detached from the Army of Italy and sent to Leopoldsdorf to form the army's extreme right flank. Grouchy had his own dragoon division and the divisions of Pully and Montbrun. To Grouchy's left was Davout's III Corps, which was advancing toward Markgrafneusiedl (Davout had Pajol's light cavalry brigade and the four infantry divisions of Friant, Gudin, Morand, and Puthod). To Davout's left was Oudinot's II Corps, consisting of the infantry divisions of Tharreau, Claparede, and Grandjean[14] and Colbert's light cavalry brigade. Oudinot was heading toward Baumersdorf. Next to Oudinot Eugene's Army of Italy with three infantry divisions, the Royal Guard, and Sahuc's light cavalry division was moving west of Baumersdorf. On Eugene's left was Bernadotte's IX Corps, consisting of two Saxon infantry divisions and a French infantry division under Dupas. Massena's IV Corps, with the infantry divisions of Boudet, Carra Saint-Cyr, Legrand, and Boudet, the cavalry division of Lasalle, and Marulaz's light cavalry brigade, faced northeast to cover the army's left. In central reserve at Raasdorf was Bessières' Cavalry Reserve Corps of three heavy cavalry divisions—Nansouty, Saint-Sulpice, and Arrighi (who had replaced Espagne)—and the Imperial Guard. The latter consisted of the Young Guard Infantry Division under Curial, the Old Guard Infantry Division under Dorsenne, and the Guard Cavalry Division under Walther. The rest of the army was still moving up.[15]

Napoleon had envisioned a coordinated attack with Davout's III Corps attacking Markgrafneusiedl, Oudinot's II Corps taking Baumersdorf, Bernadotte assaulting Wagram, and Eugene supporting Bernadotte and Oudinot. The French attack against the Russbach was to begin at 6:00 P.M., but since many of the units were still moving up and different corps had to cover unequal distances, the attack was not as coordinated as Napoleon would have liked. Davout did not get into position until it was too late, nor were all of Bernadotte's units ready. Between 6:00 and 7:00 P.M., Oudinot, Eugene, and Bernadotte attacked.

Oudinot moved against Baumersdorf with three divisions. The town was strongly held, and the portion south of the Russbach formed an effective bastion. Although the attempt to directly assault and flank the town was made, Oudinot could make no headway and was eventually thrown back.

The specific mission given the Army of Italy was to support Bernadotte and Oudinot. Eugene's army was deployed with Macdonald's corps (now containing only Lamarque's division) on the left; Grenier's corps, with Serras and Durutte, was on the right. Grenier deployed three of his infantry division's brigades (those of Moreau, Dessaix, and Valentin) in brigade columns, keeping a fourth brigade, Roussel's, in reserve. Sahuc's light cavalry division was to support Grenier, and the artillery of the Royal Guard was deployed to Grenier's right to support Oudinot against Baumersdorf. Some artillery of the Imperial Guard and Dupas' infantry division from IX Corps were sent to support Macdonald.[16]

Eugene's attack was made by echelons from the left. Macdonald crossed the Russbach, followed by Grenier, as the Army of Italy drove the defending Austrians from its banks. Roussel's brigade was committed in an effort to break through the Austrian lines. In the meantime, Dupas' division moved against the eastern face of Wagram but was held up by the Austrian defense. Eugene's army was able to advance up the slope and gain the plateau, but since none of the units on Eugene's flanks were making any headway, the Army of Italy soon became isolated and its flanks exposed. With French efforts checked against Baumersdorf and Wagram, Archduke Charles was free to concentrate his reserves against Eugene. Fifteen infantry battalions were sent against Eugene's left, and an Austrian cavalry brigade consisting of the Hess-Homburg Hussars and Vincent Light Horse charged Eugene's right. The Austrian cavalry was met by Sahuc. With night falling, and under heavy pressure and artillery fire, Eugene ordered a retreat behind the Russbach, and the withdrawal was covered by Sahuc's cavalry. At this time, Bernadotte's two Saxon divisions advanced against Wagram. The Saxons spoke German and wore white uniforms just like the Austrians, and in the darkness, they became confused and started to fire on each other. Bernadotte's attack collapsed, and his troops fell back to Aderklaa. At the same time, Sahuc's cavalry was moving back to Grenier's right. In the darkness, the French infantry mistook their horsemen for enemy cavalry and in the ensuing confusion panicked and broke, only to be rallied by Eugene, his officers, and some units of the Imperial Guard.[17]

This panic and rout has been interpreted by Petre as an example of the decline of the Napoleonic infantry: "The disorderly flight of the defeated troops of Eugene's command is notable evidence of the deterioration of many of Napoleon's young soldiers. No such panic is conceivable, under similar circumstances, in the armies of Austerlitz or Jena."[18] But this is a mistaken appreciation of the quality of the infantry, the changing nature of warfare, and the confusion of night actions. Neither Petre nor many historians after him have fully studied the campaigns of the Army of Italy in this war. Wagram was the Army of Italy's fourth major battle, and that army had engaged in eight lesser battles, several sieges, and countless skirmishes since April. Except for a minor defeat at Pordenone and a lost battle at Sacile, the troops of the Army of Italy had won many hard-fought victories and had performed magnificently in combat.

Petre wrote in the early twentieth century, prior to World War I, and before the concept of battle fatigue and modern psychological studies of the effect of combat on human behavior had been developed. Modern studies of combat units have indicated that there is a decline in morale and cohesion in proportion to the number of casualties units sustain and the amount of time they spend in operations. Prior to its commitment at Wagram, the units of the Army of Italy had lost 50 percent of their original strength since the start of the war. A typical infantry division, for example, began the war with a complement of 7,200; at Wagram, they averaged 3,600.[19] The Army of Italy actually lacked 140 line officers at the start of the war, and the missing leaders, as well as the severe losses already sustained, were bound to have a damaging effect on morale and cohesion.

What is also important is that the French were not facing the armies of 1805 or 1806. The opposition had significantly changed. Not only had the post-1805 reforms taken hold in the Austrian army, but the Austrian army at Wagram consisted of a large proportion of veteran troops. They had beaten the French at Aspern-Essling and had gained tactically in experience. The morale and cohesion of the Austrian forces were greater than before, and they could produce a greater volume of firepower. The impact of prolonged artillery fire on Eugene's forces was extremely demoralizing. Therefore, considering the uncertainty of fighting at night, the added strains of the campaigns of 1809, and the increase in firepower on the battlefield, it is little wonder that the Army of Italy temporarily panicked and ran.

The losses sustained by that army on July 5 were heavy, and the toll among the generals was great. Two division commanders, Serras and Sahuc, were wounded, and General Grenier had his hand shattered rendering him *hors de combat*.[20] Since none of the sources mention Grenier during the second day of the battle, it appears that Eugene took over direct command of Grenier's corps while Macdonald remained in command of his own.

Reinforcements joined Napoleon's army during the night. The infantry divisions of Broussier and Pacthod crossed the Danube and joined their comrades in the Army of Italy, Marmont's corps arrived and was redesignated as the XI Corps, and Wrede's division of Bavarian troops also joined the army on the Marchfeld.

Archduke Charles could be pleased with the day's results. Although he had not stopped the French from crossing the river, his forces had effectively stopped the French along the Russbach. The French army appeared to have walked into a trap. With the bulk of Napoleon's forces deployed along the Russbach their flanks seemed vulnerable for envelopment, which is what Charles planned for the next day. The right wing of his army, consisting of Kollowrat's III Corps and Klenau's VI Corps, would attack the French left with the objective of rolling them up and cutting them off from their bridges at Lobau. It was hoped that John's army would arrive and turn the French right. As a preliminary step, Rosenberg's IV Corps was to cross the Russbach and attack Davout with the intention of drawing off French reserves. Lichtenstein's Reserve Corps, Bellegarde's I Corps, and Hohenzollern's II Corps would support the attack of the right wing. What is interesting about the plan is that Charles did not include the V Corps in the operation. Napoleon would have committed every man and gun for what would be the decisive act of the battle if the situations had been reversed. However, Charles was still dominated by his pessimism and by a desire to preserve the army rather than win a decisive victory. Consequently, he kept the V Corps back on the Bisamberg to cover a retreat should that be necessary.[21]

Napoleon's objective for the following day centered around the Austrian forces behind the Russbach. His plan seemed to be that his IX and II Corps were to pin down the Austrians while Davout's III Corps turned the Austrian left. At the decisive moment, Eugene would be used to breach the Austrian line at Wagram while the rest of the army remained in reserve.[22] It is apparent that there was little concern about any threat to the French left flank, but the danger to that sector increased as part of a chain reaction involving Bernadotte and his IX Corps.

The relations between Bernadotte and Napoleon had never been warm. In a large measure, Bernadotte owed his position to the fact that his wife, Désirée, was the sister of Joseph Bonaparte's wife and so related to Napoleon by marriage. Although personally a brave man, Bernadotte's performance had been erratic, his most notorious action as a commander having been his failure to support Davout during the battle at Auerstadt in 1806. The repulse of his attack on Wagram on July 5 caused Bernadotte some embarrassment, and to salvage some of his pride, he criticized Napoleon to some officers, stating that the crossing of the Danube and subsequent battle had been mismanaged and that if he were in command, he would compel Archduke Charles to surrender by a "scientific maneuver."[23] This remark was reported to Napoleon, and considering the excitement the Emperor displayed that morning as well as the great achievement of the day, the remark seemed particularly insulting. Napoleon's famous temper flared. The incident was minor were it not connected with subsequent actions on the battlefield.

The village of Aderklaa formed a strongpoint in the French line and was to serve as a jumping-off point for the projected attack on Wagram. After the repulse from Wagram on July 5, Bernadotte's corps was to hold Aderklaa. Bernadotte felt that the position was too exposed and wanted to tie in his flanks more securely to Eugene and Massena, so sometime between 3:00 and 4:00 A.M. on July 6, Bernadotte evacuated Aderklaa on his own initiative. Bellegarde, holding Wagram, was on the alert and seized this opportunity promptly by sending Austrian troops to secure Aderklaa. It was at this time that the Austrian offensive began.

Rosenberg's IV Corps struck Davout at 4:00 A.M., and Davout had already begun an eastward movement to outflank Markgrafneusiedl when the attack came. Initially surprised, the French defended themselves well. By 6:00 A.M., Rosenberg had been thrown back behind the Russbach. Rosenberg's mission was to draw Napoleon's attention to this sector, and he succeeded in doing just that. Napoleon reinforced Davout with Arrighi's heavy cavalry division and some additional artillery. With the reinforcements came Napoleon's order for Davout to attack and carry Markgrafneusiedl, but Davout would need two hours to send part of his corps downstream to flank the position. It was at this time that Napoleon learned of Bernadotte's evacuation of Aderklaa.

An enraged Napoleon ordered Bernadotte and Massena to retake the village at all costs.[24] This order drew Massena's corps further north, and he left Boudet's infantry division to cover the approaches to Aspern and the French rear. As Bernadotte moved his troops forward to attack Aderklaa, they came under an intense artillery bombardment from Bellegarde's corps. The Saxon infantry broke and ran. Attempting to rally his men, Bernadotte galloped back to head them off and ran right into Napoleon. "Is that the scientific maneuver by which you were going to make the Archduke lay down his arms?" asked a sarcastic Napoleon. When Bernadotte tried to stammer a reply the Emperor said; "I remove you sir from the command of the army corps which you handle so badly. Withdraw at once and leave the Grand Army within twenty-four hours; a bungler like you is no good to me."[25] Napoleon then turned his back on Bernadotte; the IX Corps was out of the battle.

Meanwhile, Bellegarde's entire corps was becoming committed to the fight for Aderklaa, and Lichtenstein's Reserve Corps was compelled to keep pace with its neighbor. The Austrian grenadiers advanced to the southern side of Aderklaa to tie in with Bellegarde. With the IX Corps gone, the French position along the Russbach was threatened. To fill the gap left by the rout of IX Corps, Massena was compelled to draw his corps further north to link up with Eugene, and he still had to retake Aderklaa.

Three of Massena's infantry divisions became committed to the fighting around Aderklaa, and Nansouty's heavy cavalry division was committed to support Massena. This left the French flank and rear dangerously exposed. Only Boudet's infantry division and a light cavalry division remained to cover the French left from Breitenlee to Aspern, a distance of over two miles. Rosenberg's attack and the incidents at Aderklaa had set up the French for destruction by forcing Napoleon to concentrate on the Russbach and expose his left flank.

Major offensives developed at opposite ends of the battlefront between 9:00 and 10:00 A.M. On the French right, Davout had planned a converging attack to take Markgrafneusiedl and to destroy Rosenberg's IV Corps. The infantry divisions of Morand and Friant would attack the village directly from the south side of the Russbach while the village would be flanked downstream by the infantry divisions of Gudin and Puthod and the cavalry divisions of Arrighi, Pully, Montbrun, and Grouchy. To cover the movement and prepare the way, Davout relied on his artillery. Rosenberg had a total of 86 guns.[26] The increase in Napoleon's artillery since Aspern-Essling was a great help. Davout had 114 guns.[27]

Davout's artillery opened a thunderous bombardment, and the sixty guns of Oudinot's II Corps fired in support. The Austrian gunners responded but were at a disadvantage as they were outnumbered and caught in the crossfire of Davout's guns. By 10:00 A.M., the bulk of Rosenberg's guns had been silenced. Supported by the huge cannonade, Davout's infantry attacked, and Davout himself led the assault across the Russbach. The fighting was intense. Two Austrian generals were killed and four wounded. Davout had his horse shot out from under him, and General Gudin was wounded. The III Corps was on the verge of smashing Rosenberg when Austrian reinforcements arrived. Archduke Charles personally led up heavy cavalry from Lichtenstein's Reserve Corps and infantry from Hohenzollern's II Corps. This Austrian assault drove back the first line of Davout's men, but his reserves stopped them and drove the Austrians back. The French took full possession of Markgrafneusiedl, and Grouchy's cavalry swept around the Austrian left flank. With pressure from three sides, Rosenberg's IV Corps began to crack, and soon the entire Austrian position on the Russbach had been turned. It was now noon.

While the events of the French right were crowned with success, danger threatened on the left. Kollowrat's III Corps and Klenau's VI Corps did not move out of their bivouacs until 4:00 A.M., and by 8:00 A.M., the two Austrian corps had rolled out of the Bisamberg and were approaching Süssenbrünn, Breitenlee, and Aspern. Boudet attempted to stop Klenau by sending his artillery to enfilade the Austrian corps, but the guns were taken in flank by Klenau's cavalry and captured. Without his artillery, Boudet was forced back into the Mühlau bridgehead by 10:00 A.M. The

French were caught off balance, and their left wing was now exposed and open to envelopment.

Napoleon learned of this threat at 9:00 A.M. and ordered Massena to disengage his corps, march south to Essling, head off Klenau, and prevent him from getting to the bridges. This movement was covered by Bessières' Cavalry Reserve Corps, which launched a series of charges to keep Lichtenstein at bay as Massena marched south.

With the IV Corps marching toward Essling, a gap was created in the French line between Breitenlee and Aderklaa. Eugene, on his own initiative, had placed Macdonald's corps, now consisting of Lamarque's and Broussier's divisions, on a front oblique to his own line to halt any Austrian advance from Breitenlee;[28] the 40 guns of the Army of Italy were moved to support Macdonald as well. Napoleon personally intervened and reinforced Macdonald's front with 72 guns from the Artillery of the Imperial Guard.[29] The fire of the 112 French guns swept the area between Breitenlee and Aderklaa, halting Lichtenstein and Kollowrat in their tracks.

Klenau reached Aspern and began to advance toward Essling, but in so doing, his corps came within range of the guns on Lobau Island. The 129 cannons there poured a devastating fire into the flank of Klenau's troops and effectively checked any forward movement.

By noon on July 6, the entire fourteen-mile battlefront was aflame from the fire of a thousand guns. In many places the ripe corn growing on the Marchfeld caught fire from shells and gun wadding. Units were seen moving to avoid the flames; the wounded who could not escape were burned alive.[30] The Battle of Wagram was a battle of attrition determined by the numbers of troops and the volume of firepower. On this day, the advantage lay with the French. Charles had committed all of his troops and guns, and John's army and Reuss' corps were too far away to help. Napoleon still had over a corps in reserve, but Charles had nothing left to stop Davout from rolling up his line.

Believing that Charles had shot his bolt, Napoleon assembled a task force to pierce the Austrian center at Süssenbrunn. Macdonald's corps, positioned behind the 112-gun battery in the center, was given the mission of making a frontal assault. Macdonald was reinforced with Serras' infantry division and the light cavalry division under Sahuc (this last unit was now commanded by Gerard because Sahuc had been wounded). Macdonald would be supported on his left by Nansouty's heavy cavalry division and on his right by Walther's cavalry division of the Imperial Guard. Eugene would directly command the forces facing Wagram, the divisions of Durutte, Pacthod, and the Royal Guard.[31]

By 12:30 P.M., Massena had driven advance units from Klenau's corps away from Essling and occupied the town. The French left was secure,

BISAMBERG

Sommer
(IV) 8,000
obstng Linz.

Schustek
(V) 7,000 at (same)

REUSS

LANGENZERSDORF

HAGENBRÜNN

BISAM
HILL

Largely dispersed
along river observ-
ing crossing points

JÄGERHAUS

ARCHD. CHARLES

SAURING

STAMMERSDORF

GERASDORF

RES. LICHTENSTEIN

I BEL

STREBERSDORF

SÜSSENBRÜNN

grenadiers

LEFEBVRE
at Linz.

JEDLERSEE

LEOPOLDAU

Danube River

NUSSDORF

FLORISDORF

III KOLLOWRAT (-)

BREITENLEE

KAGRAN

MARCHFELD

HIRSCHSTATTEN

STADLAU

WÜRT.

VANDAMME

LEOPOLDSTADT
PRATER
ISLAND

VI KLENAU

ASPERN

Boudet (IV)

VIENNA LUSTAUS

ASPERN

LOBAU

Reyni
ISLAND

SCHÖNBRUNN
(2 miles)

SIMMERING

LOB-GRUND

KAISER-EBERSDORF

Schwechat

MMMMM – Austrian field works

CAMPAIGN OF 1809
WAGRAM PHASE
BATTLE OF WAGRAM (SECOND DAY)
Situation About 1030, 6 July 1809

0   1   2
SCALE OF MILES

A   B

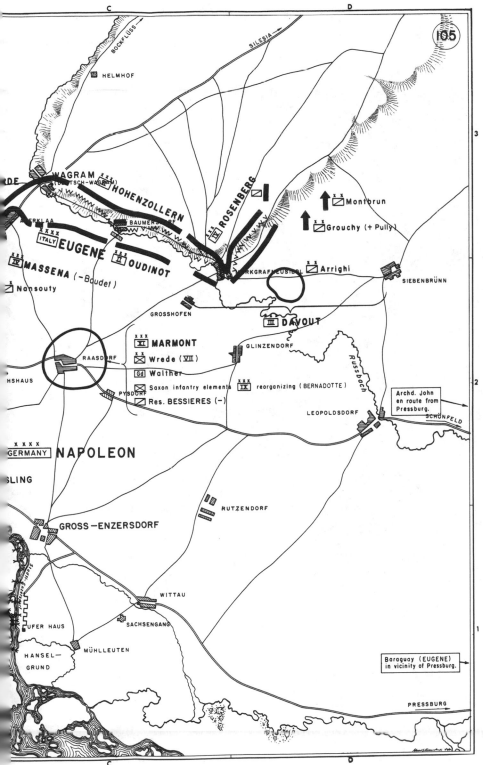

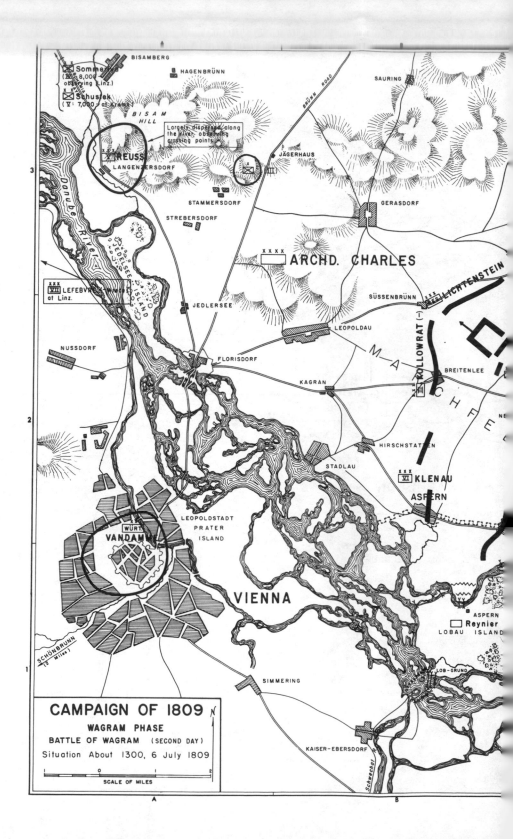

BISAMBERG

HAGENBRÜNN

SAURING

Sommexxx
(8,000 observing Linz.)

Schustek
(7,000 at Krems)

BISAM HILL

Largely dispersed along
the river, observing
crossing points

JÄGERHAUS

REUSS

LANGENZERSDORF

STAMMERSDORF

GERASDORF

STREBERSDORF

Danube River

LEFEBVRE moved
at Linz.

JEDELSEE ISLAND

JEDLERSEE

ARCHD. CHARLES

SÜSSENBRÜNN

LICHTENSTEIN

KOLLOWRAT (-)

BREITENLEE

NUSSDORF

FLORISDORF

LEOPOLDAU

KAGRAN

M·A

HIRSCHSTÄTTEN

STADLAU

KLENAU

VI

WÜRT

VANDAMME

LEOPOLDSTADT
PRATER
ISLAND

ASPERN

VIENNA

ASPERN

Reynier

LOBAU   ISLAND

SCHÖNBRUNN
(2 Miles)

LOB-GRUND

SIMMERING

KAISER-EBERSDORF

Schwechat

CAMPAIGN OF 1809

WAGRAM PHASE

BATTLE OF WAGRAM (SECOND DAY)

Situation About 1300, 6 July 1809

1   0   1   2

SCALE OF MILES

A   B

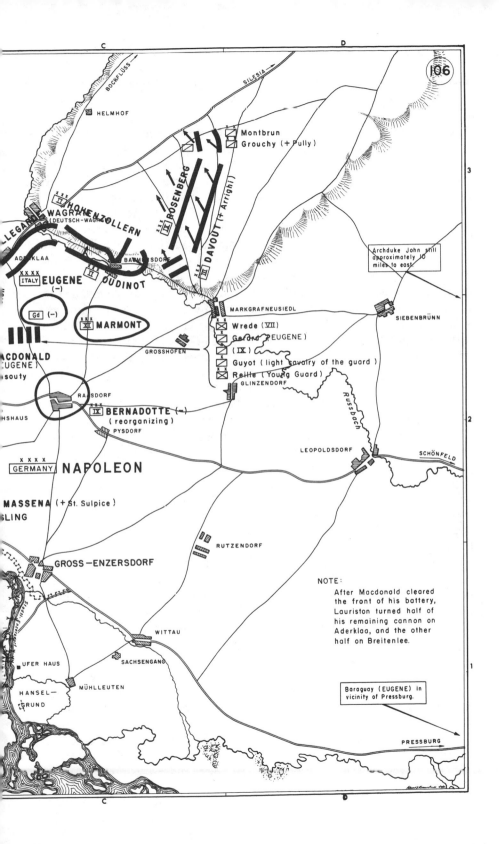

BOCKFLUSS

SILESIA

HELMHOF

Montbrun
Grouchy (+ Pully)

ROSENBERG

DAVOUT (+ Arrighi)

Archduke John still
approximately 10
miles to east.

SIEBENBRÜNN

XXX
HOHENZOLLERN

XXX
WAGRAM
(DEUTSCH-WAGRAM)

ALLEGARDE

ADERKLAA

BAUMERSDORF

XXXX
ITALY EUGENE
(−)

II
OUDINOT

Gd (−)

XXX
MARMONT

MACDONALD
(EUGENE)
Lasouty

GROSSHOFEN

MARKGRAFNEUSIEDL

Wrede (VII)
Gerard (EUGENE)
(IX)
Guyot (light cavalry of the guard)
Reille (Young Guard)

GLINZENDORF

Russbach

RAASDORF

IX
BERNADOTTE (−)
(reorganizing)

PYSDORF

HSHAUS

LEOPOLDSDORF

SCHÖNFELD

XXXX
GERMANY NAPOLEON

MASSENA (+ St. Sulpice)
SLING

RUTZENDORF

GROSS−ENZERSDORF

NOTE:
After Macdonald cleared
the front of his battery,
Lauriston turned half of
his remaining cannon on
Aderklaa, and the other
half on Breitenlee.

WITTAU

UFER HAUS

SACHSENGANG

Baraguay (EUGENE) in
vicinity of Pressburg.

HANSEL−
GRUND

MÜHLLEUTEN

PRESSBURG

and Massena turned west to attack Klenau. On the French right, Davout began to roll up Rosenberg. At this point, Napoleon ordered Macdonald to begin his attack.

The nature of the ground and the location of the enemy determine tactical deployments. The ground over which Macdonald was to attack was largely flat, his own flanks were not anchored and were covered only by cavalry, and the enemy possessed significant cavalry and artillery. The Austrian III Corps, Reserve Corps, and I Corps held the line from Breitenlee to Wagram. Macdonald's corps would be vulnerable to a converging attack, especially by cavalry. Depth in the assault was also needed to penetrate the Austrian line. For these reasons, Macdonald deployed his corps in a boxlike formation, with eight battalions deployed in two lines forming the top of the box and an additional thirteen battalions deployed in columns forming the sides and rear.[32] A more open deployment in the usual two lines of battalions would have been vulnerable to a flanking assault and would have lacked the strength to penetrate. Moreover, a wider formation would have interfered with the 112-gun artillery battery that was needed to support the attack. It was for these sound tactical reasons that Macdonald deployed his corps and not to keep the troops together because of their inferior quality, as other writers have stated.[33]

The fire from the French artillery had effectively silenced the Austrian guns when Macdonald attacked at 12:30, and the front line of Lichtenstein's corps collapsed at the impact of the assault. As expected, units of the three Austrian corps closed in on Macdonald's flanks. The Austrian envelopment of Macdonald was beaten off, but the French cavalry failed to exploit the breach. Bessières had been wounded and taken from the field, and Walther would not move without explicit orders from him. Nansouty was busy reorganizing his hard-worn division, so the Austrians were able to rally and regroup. Macdonald's advance was stalled halfway to Süssenbrunn.

Help was on the way. Napoleon ordered Wrede's division and the Young Guard to reinforce Macdonald. In addition, with Hohenzollern's flank now being threatened by the defeat of Rosenberg and the advance of Davout, Oudinot's II Corps and Marmont's XI Corps were ordered to cross the Russbach and attack Hohenzollern. Eugene, with the rest of the Army of Italy, was ordered to attack Wagram and Bellegarde's I Corps; his assault began at 2:00 P.M.[34] Rosenberg was already retreating and Hohenzollern had begun to fall back when this latest French attack began, and Bellegarde was compelled to fall back as well. With his left wing retreating and no more reserves to influence the battle, Charles ordered a general withdrawal.

Eugene took Wagram, and Macdonald took Süssenbrünn. From there the two forces converged at Gerasdorf by nightfall, where the Army of Italy was reunited under Eugene's command. Other units of Napoleon's army followed the Austrians for a while, but an immediate pursuit was impossible, for Napoleon's army had been fighting for sixteen hours and the troops were exhausted. The Austrians were not routed, and Napoleon did not know for sure until the following day that the battle was over and that the Austrians were in retreat to Bohemia.[35]

Archduke John at the head of 13,000 troops appeared three miles east of Markgrafneusiedl at 4:00 P.M., surprising some of the French troops in the area. However, the French soon stood to arms and John, realizing that the battle seemed to be over, withdrew to the east. It had taken John eighteen hours to march thirty miles from Pressburg to the battlefield. Some have blamed him for moving slowly,[36] and imply that if he had hurried, he could have arrived earlier and helped to stop Davout on the critical day. However, it actually took John about the same time to march from Pressburg to the battlefield as it took Eugene, who had the advantage of the better roads on the southern bank of the Danube.

Wagram had been a terrible battle. French casualties totaled 37,568, which broke down into 5 generals, 238 other officers, and 6,658 enlisted killed; 27 generals, 883 other officers, and 25,847 enlisted wounded; and 4,000 troops and twenty-one guns captured.[37] Included in these totals were the rather high losses incurred by the Army of Italy. Of the 20,300 troops under Eugene's direct command on July 5 and 6, 6,350 were casualties by the end of the battle,[38] and many senior officers were killed or wounded. The 112th Infantry regiment, for example, lost all of its senior officers and was being commanded by a captain who had been wounded three times.[39]

The Austrian losses totaled 41,750. This figure consists of 4 generals, 120 other officers, and 5,507 enlisted killed; 13 generals, 616 other officers, and 17,490 enlisted wounded; and 18,000 troops and nine guns captured.[40] The heaviest losses were sustained by the forces fighting behind the Russbach.[41]

Wagram was not an Austerlitz or Jena, in which the defeated army was destroyed. Charles's army was in retreat and heading to Bohemia; however, it was not a routed army and was still capable of effective resistance. Napoleon had succeeded in driving Charles to the northwest; he could now cut Charles off from Hungary and Archduke John. Moving an army of such immense size across the river and depositing it on the north bank ready to engage in a titanic struggle was an amazing achievement. So why did Napoleon fail to destroy Charles' army? The theory that Napoleon's infantry was in decline due to the high ratio of conscripts in its ranks is nonsense. The performance of the infantry in the French II, III, and IV Corps was magnificent, as was that of the infantry in Eugene's Army of Italy. The lone panic on the evening of July 5 has already been explained.

The rout in IX Corps was by Saxon infantry whose tactics and leadership were antiquated. The real reasons that this battle seemed so tactically different from Austerlitz, Jena, or Friedland were the improved organization and performance of the Austrian army and the dynamic created by relatively evenly matched forces.

The army corps structure (or *columns* as they were now called in the Austrian army), which enhanced command and resiliency, mitigated against any disintegration, just as it did for the French (except perhaps for the IX Corps). It was the corps structure on both sides that facilitated the employment of huge masses of artillery and ostensibly turned the Battle of Wagram into an artillery duel. The fire of those massed guns effectively halted any forward tactical movement, as the fire of the French batteries in the center and on Lobau attested. The firepower delivered by the rival corps made the battlefield more deadly and prevented the type of tactical breakthrough that had occurred in the past. The increased lethality on the battlefield helps explain the high number of officer casualties in the French army and the disparity in losses between the French and Austrian officers.

Petre argues that the high number of French officer casualties was because "the French officers had to sacrifice themselves more freely on account of the quality of their troops."[42] But Petre errs in thinking that the quality of the infantry had declined and fails to realize that in most of the tactical engagements in this battle, the French did more of the attacking and the Austrians did more of the defending. Attacking places a greater premium on leadership than defense does. The burdens on the leaders increased because of the enhanced firepower from the enemy formations. More officers had to lead attacks to get through the beaten zone. In such situations it is therefore natural that the officer casualties of the attacking force will be proportionately higher. The only maneuver that succeeded in this battle was that of turning a flank, and this Davout was able to do because he used his superior artillery to suppress Rosenberg's guns and was thus able to overlap the Austrian flank. This was a battle largely determined by attrition, and Napoleon, with greater resources in men and materiel, won for that reason.

Unlike many previous Napoleonic battles, it was the opponent who seemed to have the initiative most of the time. Napoleon surprised Charles by crossing the river, yet from the time the fighting began on the Russbach until Klenau and Kollowrat were stopped at midday on July 6, it was Charles who shaped the battlefield. The battle area was too huge for effective French reconnaissance, and Napoleon evidently did not know of the location of the Austrian III Corps until 9:00 A.M. on July 6. Rosenberg's initial attack and the actions at Aderklaa drew French troops and Napoleon's attention to the Russbach front. Napoleon's fixation on this area exposed his army to envelopment, and it was French firepower and the superb handling of the IV Corps that saved the left wing. One can only wonder whether the Austrians would have cracked the French left if Charles had ordered Reuss' V Corps to move along with Kollowrat's and Klenau's troops. Its absence rendered Charles' pessimism about the course of the battle both self-fulfilling and self-defeating.

Macdonald, Oudinot, and Marmont were made marshals by the Emperor after the battle. Macdonald's battlefield promotion clouded the issue of the command relationship in the Army of Italy, as he used the promotion to play up his role as the de facto commander in chief.[43] Only a corps commander in the Army of Italy, Macdonald's prominent role on the second day of the battle was due to the mere chance that his corps was on the left, where significant action occurred, rather than the right, and because Grenier had been wounded. During the crisis of July 6, Macdonald's location on the battlefield meant that he temporarily became a task force commander under Napoleon's direct command. Once his singular mission was completed, Macdonald was returned to Eugene's command.

As a commander, Eugene did a fine job at Wagram. Forces beyond his control drove his troops back behind the Russbach on July 5, and he displayed great initiative in the positioning of Macdonald's corps to cover the gap in the line on July 6. Eugene did all that was required in this battle.

Active operations resumed late on July 7. Napoleon had lost contact with Charles' army but knew it was heading north. Once again, the Emperor was concerned about the security of his flanks and his lines of communication north and south of the Danube. The Armies of Italy and Germany were given separate missions.

Eugene was given the task of defending the left and rear of Napoleon's army. The Viceroy's army consisted of the V, VI, VIII, IX, and XII Corps, the IX Corps now commanded by General Reynier. Keeping most of his forces north of the Danube, Eugene marched east looking for the army of Archduke John and had gotten as far as Stamphen by July 13, when an armistice was declared.

Napoleon took his army north to chase Archduke Charles. He sent Massena into Bohemia and Davout and Marmont into Moravia to look for Charles' army, keeping Oudinot and the Imperial Guard in reserve. Massena located the enemy army and fought with its rear guard on July 8. Learning that Charles was heading for Znaim behind the Thaya River, Napoleon directed his army there. Charles had crossed that river and reached Znaim on July 10 when he was attacked by Marmont, who had previously crossed the river. Marmont was pushed back, but the combat gave time for Massena to arrive.

Archduke Charles had evidently had enough. He was becoming effectively isolated from Hungary by Napoleon's forces, and Jerome's X Corps was advancing against northern Bohemia. Charles still had an army, but if he lost one more battle it could mean the end of the Hapsburg state. On his own initiative, Charles offered Napoleon an armistice, which was accepted on July 12 and signed the next day, effectively ending the war. The armistice terms stipulated that the Austrian forces were to evacuate the Tyrol, Graz, Croatia, Bohemia, and Moravia and were to withdraw behind the March and Raab rivers until a final peace treaty was made. The armistice terms enraged Emperor Francis, who dismissed Charles on July 18.

Francis hoped to use the respite to rebuild his armies, because he knew that British help was coming. On July 28, Britain mounted the largest amphibious operation of the Napoleonic Wars. Two hundred and sixty-six ships including forty-four sail-of-the-line and twenty-two frigates brought an army of 44,000 troops to the Scheldt estuary.[44] The objective was to capture Flushing and Antwerp, draw off Napoleon's armies, and ignite a revolt against the French in northern Europe. If this operation had occurred in April or May, there may have been more chance for success. Now it was too late, and the British land and sea commanders, respectively, the Earl of Chatham and Richard Strachan, bungled their jobs. By August, the British forces were effectively blockaded on Walcheren Island by second- and even third-rate French troops. Napoleon knew that Walcheren was a pestilential island and merely ordered the French to contain the British, knowing that malaria and typhus would infect the British army. Eventually 12,000 British troops were incapacitated by "Walcheren Fever." By September, it became clear that the British expedition was a failure, and they evacuated in December.

It was the landing at Walcheren that prompted the myth about the decline in quality of the French infantry and its compensatory increase in artillery support. On August 18, Napoleon wrote to his Minister of War, General Clarke, about the French forces blockading Walcheren. Knowing that these troops consisted mostly of raw conscripts and national guards, Napoleon ordered them to be well supported with artillery, because "the poorer the troops, the more artillery it needs."[45] Petre took this statement out of context and applied it to the situation and troops at Wagram.[46] He and others used it to build the theme that the mass formations, the increased use of artillery, and the high rates of officer casualties were all due to the poor quality of the infantry at Wagram. This is not true. Napoleon wrote about third-rate troops at Walcheren; there is no evidence in the *Correspondance* to support the idea that he thought the first-line troops at Wagram had in any way declined. The conscripts in the French Armies of Germany and Italy were worthy successors to those of 1805–1807, and the increases in artillery and officer casualties were due to changes made by the Austrian army and their corresponding effect on the French army and the dynamics of war.

The British failure at Walcheren became evident to Emperor Francis by September. The failure at Walcheren and Napoleon's victory at Wagram ensured that no European power would help Austria. The resources at Napoleon's command were far greater than Francis' were, so it would be hopeless to renew the war.

The peace treaty was signed at Schönbrunn on October 14, 1809. Austria was to cede to Bavaria the provinces of Salzburg and Inn-Viertel. It was to deliver to Napoleon, as Emperor of the French and King of Italy, parts of Friaul, Carniola, Carinthia, the port of Trieste, and all of Croatia and Dalmatia south of the Save River. Austria's Polish territory was divided between the Duchy of Warsaw and Russia. The loss of territory included the loss of 3 million of Austria's 16 million inhabitants. In addition, Austria was to pay a war indemnity of 85 million francs and limit the size of its army to 150,000 men.[47]

The answer to the question asked at the beginning of this book—Was Austria still a great power?—must be yes, since it engaged and taxed the resources of the Napoleonic Empire. Was Austria a great power after the War of 1809? It was certainly no longer a great power of the first rank. The loss of territory, the war indemnity, and the limitation on the size of the army ensured that Austria alone would not be able to engage Napoleon again in central Europe. Nor could Austria have a completely independent foreign policy after the Peace of Schönbrunn. The Austrian Empire now was clearly a European power of the second rank. Internal integrity remained, however, and the Austrian Hapsburgs did not suffer the same fate as the Spanish Bourbons. What Austria won on the battlefields of Sacile, Aspern-Essling, and Wagram was Napoleon's respect for its military prowess.

This is illustrated by the terms of the peace treaty. Austria in 1809 was not treated like Prussia in 1807. Napoleon was justified in claiming compensation for Austria's unprovoked attack, yet the terms were not vengeful. Napoleon wanted peace in central Europe. He still faced a war with Britain and in Iberia, and he hoped the current and comparatively easy terms of the treaty would bring peace in central Europe. Moreover, the Russians would not countenance the destruction of the Hapsburg state, leaving themselves alone in Europe to face Napoleon. The continued existence of the Austrian army after the Battle of Wagram was a factor in the relatively easy peace terms; Napoleon's respect for Austria was shown after the treaty was signed.

Seeking both legitimacy and security, Napoleon sought a marriage alliance with the Hapsburgs the following year. At a meeting of the Imperial Council in 1810, the Austrian marriage was discussed. When Lacuee de

Cessac opposed the marriage, stating that "Austria is no longer a great power," Napoleon replied, "It is easy to see that you were not at Wagram."[48] Evidently, Napoleon thought the alliance worthwhile, since there was no point in having a marriage alliance with a weak power. Napoleon no doubt lost confidence in his erstwhile ally, Alexander of Russia, because of his performance during the recent war. Russia's minimal support in 1809 started the process that led to war with France in 1812, and Napoleon turned to an Austrian alliance in an attempt to overawe Russia. For these reasons, and the fact that Josephine was unable to produce an heir, Napoleon divorced her late in 1809 and married Maria Louise of Austria the following year.

Eugene, understanding that the divorce was for reasons of state, supported his stepfather's decision, even though he was distraught over the affair. He offered to resign from his positions to stay with his mother, but Napoleon would not allow it. He promised that Eugene would retain his confidence and affection, which Eugene returned in kind. Although he never got the Italian throne, he not only kept his position as Viceroy of Italy but was also given greater responsibilities.

At the close of the Austrian war, Eugene conducted a counterinsurgent campaign that finally ended the Tyrolean revolt. In 1812, he commanded one of the three field armies that invaded Russia and was given command of the remnants of the *Grande Armée* during the last stages of the retreat. During the late winter and early spring of 1813, Eugene strove to hold Germany for Napoleon, and from the spring of 1813 to March 1814, he defended the Kingdom of Italy until Napoleon's first abdication. He had been offered the crown of Italy if he would defect, but Eugene refused and remained loyal to his one-time stepfather to the end.

CHAPTER TWELVE

The Emergence of Modern War

Although Napoleon did not know it at the time, the Franco-Austrian War of 1809 was the last war he would win. Napoleon had set forces in motion that he would be unable to control or, worse for him, unable to understand. He had created a modern army and had helped establish the operational campaign. In his own mind, the operational campaign ended with the decisive battle. However, the evolution of the conduct of the operational campaign was not complete. Napoleon's decisive victories were possible only against the obsolete armies of the *ancien régime*, and this mismatched relationship brings us to the verge of the modern campaign. The dynamic of warfare did not become fully modern until other armies modernized, creating a symmetrical operational dynamic that ended the decisive battle and replaced it with a series of engagements whose strategic outcome was cumulative. The full emergence of modern war began in 1809.

I began this book with a definition of modern war that considered a war modern when there was a strategic war plan that effectively integrated the various theaters of operations, when it was marked by the fullest mobilization of the resources of the state, and when operational campaigns were used to achieve strategic objectives in the various theaters of operations. In addition, those operational campaigns were characterized by the use of opposing symmetrical armies raised by conscription, organized into army corps, maneuvered in a distributed fashion so that tactical engagements are sequenced and often simultaneous, with decentralized command and control, and a common doctrine. Victory is achieved by the cumulative effect of tactical engagements and operational campaigns. Does the war of 1809 meet the criteria?

Strategically, both sides had integrated war plans that encompassed different theaters of operations. Both sides realized that the German theater of operations was the most important, but the other theaters, especially the Italian, also had roles to play.

Austria mobilized the resources of the state on a scale greater than ever before in its history. Conscription was introduced within the provinces of Austria and Bohemia, and appeals were made to German nationalism. Austria conducted simultaneous offensives in all theaters of operations. Its greatest army, totaling eight corps, was sent to the major theater of operations in Germany, two corps were sent to Italy, and one to Poland. Austria's strategy was to surprise and attack the French simultaneously in all theaters and gain quick victories in all. This was perhaps Austria's greatest strategic mistake. Rather than expend precious resources in peripheral areas that produced little strategic gain, Austria should have remained on the defensive everywhere except in the primary theater of operations. Along the Danube, Austria had a chance of defeating the main French army and gaining control of southern Germany and the upper Danube. But by conducting offensive operations in the secondary theater, Italy and Dalmatia, Austria effectively threw away an important strategic asset, the Army of Inner-Austria. After being beaten in Italy and Dalmatia by Eugene's numerically superior army, John's army could neither defend the southern frontiers of the Austrian Empire nor prevent the junction of Napoleon's Armies of Germany and Italy. If John's army had not invaded Italy and Dalmatia and had instead remained holed up in some of the most defensible country in the world, Eugene's army could have been either neutralized or destroyed trying to break through the Alpine defenses.

At this point, one must speculate on what might have occurred. Even though Napoleon initially defeated Archduke Charles' army in April, he still needed the Army of Italy. Without Eugene's army, Napoleon would have had 40,000 fewer troops and 100 fewer guns for operations against Charles, reducing the rival strength of the main armies to numerical equality. Since Napoleon had to use the corps of Lefebvre and Baraguey d'Hilliers to secure his flank and rear when he crossed the Danube, Eugene's army was critical to the success at Wagram. Had Eugene been defeated in the Alps, John could have sent one or more corps against Napoleon's flank and rear, creating such a threat that Napoleon would have been forced to detach even more troops to deal with him or abandon the offensive north of the Danube entirely. The Army of Italy provided the margin of victory at the Battle of Wagram. The junction of the two armies, Napoleon's and Eugene's, and the convergence of two lines of operations ensured victory for the eagles of France.

France's strategy was better than Austria's. Having chosen to put the onus for causing the war on Austria's shoulders to trigger the defensive alliance with Russia, Napoleon gave his armies enough resources to absorb the initial Austrian onslaught, throw the enemy back, and pass over to the counteroffensive. Once the initiative had been regained, Napoleon aimed for simultaneous advances by all of his armies toward the heart of the enemy's country and the destruction of the main enemy army.

By adopting the army corps system, the French had been able to deploy and maneuver their armies across broad fronts. So long as their opponents failed to adopt a similar system, the French could easily concentrate against enemy forces or surround them. However, once the Austrians adopted the corps system, they, too, could deploy and maneuver armies across broad fronts, which altered the dynamics of war. Since the Austrians could deploy and maneuver broadly, the French were compelled to do likewise and so were prevented from concentrating against single targets as in the past. In Germany and Italy, the opposing armies advanced along fronts a hundred miles in length, fighting multiple related engagements. The clashes from Thann through Ratisbon in Germany in April, Eugene's invasion of Austria in May, and the operations along the Danube in June and July are the most significant and best examples of sequential and simultaneous battles. The true operational campaign emerged. Engagements were fought in the depth of the theater of operations as well as across its length. For example, the uprising in the Tyrol threatened the rear of both French armies in Germany and Italy, and the movements of the divisions of Chasteler and Jellachich were always a major concern to Napoleon and Eugene.

Besides enabling the armies to operate in breadth and depth, the corps system enhanced operational resilience and tactical power and prevented the collapse of armies after tactical defeats. At worst, only a segment of an army could collapse; the greater unit articulation ensured that reserves and a method of control existed that could mitigate against disaster. The corps system also improved the coordination of the combat arms. In particular, this allowed a greater amount of artillery to be used, and consequently, the amount of firepower increased, and battlefields became more deadly. It became more difficult for the tactical offense to overcome the fire of the defense. Attrition began to characterize the battlefield. Although both sides sought decisive battles in the old manner, neither side could win one. None of the major battles, Thann-Ratisbon, Sacile, Piave, Aspern-Essling, Raab, or Wagram, resulted in decisive victories. Napoleon's victory and Austria's defeat were the cumulative result of battles and campaigns in which the side with the greatest resources won.

Corps operating across broad fronts produce a particular type of mobile operations, which in turn places a premium on decentralized command and control. Moving and fighting on broad fronts creates confusion and uncertainty for the army commanders, since they cannot be everywhere at once. Victory and defeat are determined by the tactical abilities of subordinate commanders, who act on their own initiative according to flexible, mission-type orders. This was the situation for the operations on the Danube in April, when the initiative shown by the corps commanders, in particular Davout, Lannes, and Massena, was significant in stopping the Austrians and throwing them back. The same situation existed in the Army of Italy. Marmont, Macdonald, and Grouchy were all given broad missions and great independence to accomplish them. Marmont was to secure Dalmatia and drive into Croatia. Macdonald was to clear the Friaul, take Laibach, and advance into Carniola in cooperation with Marmont. Grouchy was to move down the Drave and support the advance of Macdonald. All three accomplished their missions. The French, with more practice, had a distinct advantage in this type of warfare. Napoleon accepted the confusion of this type of campaign as normal; so did Eugene, although he was a bit shaky at the start of the war. The acceptance and practice of decentralized command and control gave the French a tremendous advantage in this war.

Archduke Charles was not intellectually comfortable with decentralization, and the uncertainty of distributed operations paralyzed him. He could not understand this type of warfare and fell back on eighteenth-century methods of centralized command and control after the battles in April. Charles' inability to function in modern operations helped deepen the pessimism that he had about his army's abilities and the possibility of winning the war. We will never know if a bolder commander could have conducted a modern offensive operation against Napoleon after Aspern-Essling. Archduke Charles did not have the mind or the heart for such an attempt, and this was his greatest failure as a commander.

Both sides seemed to have some sort of operational doctrine. The Austrians had published a manual for the conduct of war above the tactical level, *The Fundamentals of the Higher Art of War for Generals of the Austrian Army*, and just how deeply this doctrine had penetrated into the Austrian officer corps is uncertain. The issue for the Austrians was to bridge the gap between theory and execution.

The French example presents a more complex picture in respect to doctrine. An examination of the operations of the Napoleonic armies in Germany and Italy reveals a pattern of common operational and tactical practices. Eugene's desire to envelop John's army on the Adige and his plan to destroy it piecemeal in Hungary indicate that Eugene had learned the operational themes of Napoleonic warfare. Even more decisive in proving this commonality is how the organization of the Army of Italy so closely mirrored that of Napoleon's army. The division of Eugene's army into formations designated left, right, center, and reserve, all kept within supporting distance of each other, matches the deployment of Napoleon's army. Eugene's approach to his Alpine campaign was to march his army dispersed across a broad front. In so doing he followed the principles of Pierre Bourcet and Napoleon himself. If one examines Napoleon's campaign into the Alps in 1797, one discovers that Eugene's deployment of his army and his choice of routes are almost an exact copy. In the tactical conduct of Eugene's battles, one sees the three major components of Napoleonic battle—the *masse primaire*, *masse de manoeuvre*, and *masse de rupture*—used at the battles on the Piave and at Raab. Such similarities cannot be by happenstance; rather, they are by design.

If one accepts this premise, then it is fair to say that an operational and tactical doctrine existed in the French army, but it is a doctrine of an early nineteenth-century type. Later, it became common practice to disseminate doctrine through published manuals and via a hierarchy of staff and war colleges. Napoleonic doctrine, on the other hand, was spread informally through the French army. Officers who chose to improve themselves professionally read and studied on their own, reading the same works that Napoleon read—for instance, the writings of Guibert and Bourcet. This was different from the formal Prussian practice of providing education and training for staff officers, yet it is hard to deny the existence of informal education in Napoleon's army.

In the French army, doctrine was also spread by practice. Most of Napoleon's marshals were generals in their own right at the time Napoleon came to power. It was not possible to send these men back to school in a formal sense, and it would have been resented at the very least. So, Napoleon's method of education was situational, indirect, and cumulative over a long period of time. Many of Napoleon's commanders had served with him since as early as 1796, and anyone who has been around soldiers knows how they like to talk about their profession and to exchange ideas. The image of Napoleon that emerges from his correspondence and in the memoirs of the period is that of an avid enthusiast rather than a tight-lipped introvert. After working with the same men for ten or more years, it is perfectly natural that an understanding developed between Napoleon and his commanders about the conduct of war. Massena, Davout, and Lannes were not automatons, nor was Eugene de Beauharnais. In addition, there is Napoleon's correspondence, from which other people have distilled the essence of Napoleon's methods of warfare.[1] If they could do so, then could not the commanders who were the recipients of these letters?

Napoleon clearly explained to Berthier what he wanted done on the Danube in April: a concentration against a frontal offensive—or a *manoeuvre sur les derrières*— should the Austrians head for the Main. The Army of Germany's left, right, center, and reserve were placed in mutually supporting distance from each other to maneuver according to circumstances. Berthier's mistakes in deployment were made because he did not understand the spirit of Napoleon's instructions, not because he was uninformed about what Napoleon wanted. Marshal Davout had a clear understanding of what was to be done but was compelled to obey the orders of Berthier until Napoleon arrived. Napoleon's instructions to Eugene and Marmont are a further example of the length that Napoleon would go to convey his operational concepts. The plans for a defensive campaign in Italy were well spelled out. The design for offensive operations stressed flexibility but provided direction and objectives. Familiar operational themes keep emerging in all of Napoleon's instructions: divide the enemy to defeat him piecemeal or destroy him by envelopment; disperse to compel the enemy to divide, then concentrate faster than he can in order to destroy him. Prince Eugene, for example, was made to fully understand the role of his operation in Hungary and its place in a broader strategic scheme.

There is a good deal of written evidence to prove the existence of a tutor-student relationship between Napoleon and Eugene. In part, the geographic distance separating the Emperor and his Viceroy meant that Napoleon had to write more frequently and in greater detail to Eugene. Moreover, Eugene was younger and less experienced than his marshals, and Napoleon could treat his stepson differently from the other commanders. Napoleon felt that in Eugene's case, fuller and more detailed instructions were necessary. In 1809, since Napoleon had worked more closely with the commanders of the Army of Germany and was in personal command shortly after war broke out, there is less of a "paper trail" to examine. Many conversations and verbal orders were not recorded.

The question arises, Why did some of Napoleon's marshals not perform as well in independent command in later campaigns? The answer is not in a lack of training, education, or doctrine but, rather, in human nature. No teacher, staff college, or war college can expect 100 percent perfection from its graduates, and the staff and war colleges developed in the late nineteenth and twentieth centuries have produced more failures than successes in the art of war. The successes and failures of Napoleon's commanders should be held to the same standard.

Other questions were raised in Chapter 1 about certain orthodoxies regarding the history of the 1809 war. Paramount in the usual interpretations is the idea that Napoleon's abilities as an army commander were in "decline."[2] One needs to be careful about ascribing victory and defeat only to certain individuals rather than looking for more complex factors. Many people have argued that the decline of Napoleon's genius was the reason for mistakes made during operations on the Danube in April (Thann-Ratisbon) and at Aspern-Essling in May. Perfection in war is impossible; confusion and uncertainty are inherent in its very nature. One will never understand war until one examines it in the relational context of opposing forces. If Napoleon seemed like a genius in his previous campaigns, he did so only at a given time against certain types of opponents. The relational context of warfare in 1809 was different than previous wars. The nature of the enemy had changed, which in turn changed the dynamics of warfare. Our perception of Napoleon and the others involved in that war must change as well.

If Napoleon can be faulted for making mistakes in 1809, it is for his inability to fully comprehend the changing nature of war. Until Aspern-Essling, he underestimated his opponents, which was understandable given the past performance of their army. (Even Archduke Charles had little confidence in his army.) After Aspern-Essling, Napoleon realized the tactical disadvantage he faced regarding artillery and corrected the problem for Wagram. But Napoleon could not project how the increased tactical lethality and the other changes made by the Austrian army would alter the broader conduct of war.

Real peace in Europe was not at hand, in spite of Napoleon's victory in this war and subsequent treatment of Austria. Great Britain would not accept French hegemony on the continent, and so the war between France and Britain went on. Napoleon was incapable of making the political decisions that might have ended the insurrection in Spain, so that situation persisted as well. And the strain of Napoleon's Continental System continued to be a burden on Europe.

In spite of the Austrian marriage, Francis of Austria, Alexander of Russia, and Frederick William of Prussia were still hostile to the Napoleonic regime and everything it stood for. In the final analysis, Napoleon's position in Europe rested upon the power of France and its armies.

The modernization of the European armies continued, and the qualitative advantage held by the French over the armies of Russia and Prussia diminished.[3] The Russians had adopted the army corps system by the time Napoleon and Tsar Alexander went to war against each other in 1812. Napoleon invaded Russia in 1812 with a total force of 677,000 troops and 1,393 guns.[4] His force was divided into three field armies, and the operational front extended for hundreds of miles. The Russian field armies also were organized into corps and managed to avoid encirclement. Napoleon, seeking another Austerlitz, outdistanced his logistical support and was drawn into the depths of Russia. Eventually, the Russians made a stand at Borodino, where the opposing forces—130,000 French and allies with 587 guns and 120,000 Russians with 640 guns—fought all day in an attritional battle of monumental proportions.[5] There was no breakthrough on either side. This battle, like Wagram, was dominated by the fire of the guns. Essentially the battle was a draw, although the French were left in possession of the field. The modernization of the Russian army played a part in Napoleon's eventual defeat.

With the loss of his army in Russia in 1812, Napoleon lost the power to overawe Prussia and Austria, who reentered the lists against him in alliance with Britain and Russia. In the wars of 1813–1815, victory went to the side with the greatest strategic resources.

The anti-French coalition in 1813 was greater than France. Napoleon kept hoping he could redress the strategic balance against him by a single decisive victory, but in the context of the Napoleonic Wars, that day was over.

The central European campaign of 1813 compares favorably to the later campaign of 1866. The Prussian army, which had been so badly defeated in 1806, rose from the ashes. The Prussians had thoroughly transformed their state and army. Conscription had been introduced, and they had established an expansible army. Thus, they entered the field in 1813 with 272,000 troops by mobilizing regulars, trained reserves, and *Landwehr*. They had also adopted the army corps system.

The forces raised by August 1813 were huge. The Allied armies, which included Sweden's, totaled 800,000, of which 500,000 and 1,380 guns constituted the frontline forces. Napoleon was able to mobilize 700,000, of which 400,000 troops and 1,284 guns were in the front lines.[6] The Allied forces were divided into three field armies designated the Armies of the North, Silesia, and Bohemia. Later a fourth army, Poland, joined the Allies. These armies operated on a strategic arc 350 miles in length from Brandenburg through Silesia to Bohemia. Every army was organized into corps. Napoleon, with a smaller army, was based in a central position in Saxony. The French attempted to destroy the surrounding armies piecemeal, but in spite of victories, such as at Dresden, the structural resilience of the field armies provided by the corps system and the huge volume of artillery fire precluded any decisive defeat as long as they were not surrounded and could maneuver away. It was Napoleon's army that was worn down by shifting from one front to the next.

With the arrival of the Army of Poland, the Allies chose to make a concentric attack against Napoleon and his lines of communication running through Leipzig. (The Allied armies were practicing Moltkian warfare but without Moltke.) At the same time, the French Emperor decided to concentrate his army at Leipzig in the hope of smashing the separate armies as they moved against him. Napoleon, who was considered the god of battles, was overthrown by the dynamics of warfare that he had created but did not understand. As in 1809 and 1812, the armies were too large, resilient, and powerful to be destroyed in a single battle. Napoleon was able to push Prince Schwarzenberg's Army of Bohemia back in the early phases of the Battle of Leipzig, but that was all. In reality, Schwarzenberg was able to pin down Napoleon while the other Allied armies closed in on the French.

The battle at Leipzig, fought October 16–18, 1813, was the greatest battle fought in the nineteenth century. Napoleon's army, with under 200,000 men and 799 guns, engaged an Allied host that totaled 361,000 men and 1,456 guns.[7] The battlefront extended for sixteen miles. This was a great battle of attrition in which the French were driven into a pocket, but they had a line of retreat open and escaped. The French sustained 73,000 casualties while the Allies lost 54,000,[8] but this was no Austerlitz. The Allies were too exhausted from this great bloodletting to pursue Napoleon, and his army escaped to fight into the spring of 1814. Leipzig was a strategic turning point in the course of a protracted war.

It can be argued that the last battle of the Napoleonic Wars, Waterloo, was a decisive battle, but the perspective changes if one looks at the entire campaign. The Waterloo campaign, conducted June 15–18, 1815, consisted of continuous and sequential operations. Battles were fought at Charleroi on June 15, Quatre-Bras and Ligny on June 16, and Waterloo and Wavre on on June 18. Throughout the theater of operations, the French were outnumbered two to one. Napoleon went into Belgium trying to re-create his Italian victories, which had been won on interior lines, or to win Austerlitz and Jena again. He beat the Prussians at Ligny, inflicting 28,000 casualties on an army that numbered close to 100,000 men. In the conditions of 1805 or 1806, this would have been enough to cripple an army of the *ancien régime*. But that type of army was long gone. The corps structure, an effective staff system, and national determination ensured that the Prussian army would return to the fight, and the Prussians were able to envelop Napoleon's army at Waterloo. (The envelopment by two or more armies was a characteristic of operations practiced by von Moltke in 1866 and 1870.) Fighting on two separate fronts late on the afternoon of June 18, the French army collapsed. If one examines the casualties for the entire Waterloo campaign, one sees that the French lost 63,000 while the Allies lost 61,000.[9] From this perspective, the campaign was determined by attrition and fits the pattern of the campaigns fought since 1809.

Neither Napoleon, the Allied commanders who fought him, nor the theorists Clausewitz and Jomini were able to articulate the way warfare had changed since 1809. For them, the Napoleonic Wars formed a unitary block rather than two separate periods characterized by different relational dynamics. The failure to understand this difference, brought about in part by a fascination with Napoleon's earlier triumphs, mistakenly influenced military thought well into the nineteenth century. Clausewitz, Jomini, and many generals in the American Civil War and in Europe were convinced that another Austerlitz was still possible. It was left to Moltke to articulate and identify what was actually practiced in the second half of the Napoleonic Wars.

Those wars were fought with a military technology based on smoothbore muzzle loaders and muscle power. By the middle of the nineteenth century, steam-powered railroads, the electric telegraph, and rifled weapons were commonplace. It has been argued that the advent of this technology significantly altered warfare. It did indeed, but not uniformly, and consequently, significant characteristics of the Napoleonic Wars of 1809–1815 still remained in the middle of the century.

The American Civil War witnessed the use of the new technology. The use of railroads and the telegraph primarily affected strategic deployment, logistics, command, control, and intelligence; once the armies left the railheads and telegraph lines, they moved and were supplied in the same manner as Napoleon's armies. All contact was lost with Sherman's army when it set out on its march from Atlanta to the sea, for example. The Civil War armies were organized, like the Napoleonic armies, into army corps. The size of the armies of the Civil War was actually smaller than those of the second half of the Napoleonic Wars. The Union's Army of the Potomac peaked at a numerical strength of 150,000 while the size of the Confederate Army of Northern Virginia hardly ever surpassed 80,000. Battlefronts matched those of the Napoleonic Wars. For example, 90,000 Union troops with 220 guns and 70,000 Confederates with 172 guns faced each other on a six-mile front at Gettysburg.[10]

A major theme in Civil War history concerns the tactical deadlock wrought by the combination of infantry rifles and entrenchments. This combination rendered frontal assaults suicidal and ensured that decisive victories were impossible, save by complete encirclement. Yet this was a characteristic of the Napoleonic Wars from 1809. There was no break-through at Wagram, Borodino, Leutzen, Bautzen, Dresden, Leipzig, or Waterloo. In spite of the increased range of rifled muskets, engagement ranges in the Civil War were often not at extreme range but at Napoleonic distances.[11] The rifled weapons added to the power of the tactical defense that had been established by 1809. The resilience of the armies of the Civil War was provided by the corps structure. The belief in the Austerlitz type of victory infected Lee as much as it did Napoleon, with the same disastrous results. The symmetrical relationship between the Civil War armies ensured that the American Civil War, like the later Napoleonic Wars, was a contest based on attrition.

The wars of German unification were of short duration, yet there were many similarities with the post-1809 period. The numbers of the forces mobilized for the Austro-Prussian War of 1866 matched those of the Napoleonic Wars. Austria raised 528,000 men, of which 320,000 went to the armies in Bohemia and Italy. The Prussians mobilized 355,000, of which 254,000 were used to invade Saxony and Austria. The Prussians deployed three field armies, all organized into corps, along a 250-mile front at the start of the campaign, matching the length of the operational front of 1813. The difference between 1813 and 1866 was that the Prussian armies were deployed by railroad and directed by telegraph. But once these armies left the railhead, they marched and were supplied as in Napoleon's day. There were battles fought along the frontier at the outset of the campaign of 1866 as in April 1809. The Prussian armies converged on the Austrians at Königgrätz as the Allies did against Napoleon at Leipzig, and the total numerical strength of the armies at Königgrätz was actually less than at Leipzig. The Austrian and Saxon armies totaled 206,000 with 770 guns and faced 220,984 Prussians with 792 guns along a ten mile front. König-grätz, like Leipzig, was characterized by an attempted envelopment. The defeated Austro-Saxon army had an open line of retreat for escape, as did the French at Leipzig. The Austro-Saxons suffered 44,000 casualties while the Prussians lost 9,000. Königgrätz was not an Austerlitz. The Austrian army remained intact, and the Hapsburg Empire was still capable of resistance. A political settlement ended what proved to be a limited war.[12]

The size of the forces mobilized for the Franco-Prussian War of 1870–1871 was comparable to the second half of the Napoleonic period. The Prussians and their German allies mobilized 850,000, of which 309,000 went to the frontline armies. The French Second Empire raised 567,000 troops, of which 200,000 went to the frontline armies,[13] and many more troops were raised by the French after Sedan. The opening stages of the war, referred to as the Battle of the Frontiers, matched those of 1809 and 1866. Rival forces fought each other over broad fronts in parallel and sequential engagements. The Prussians managed to maneuver between the French armies and surround them at Sedan and Metz, compelling their surrender. These victories were decisive in that the French armies were completely trapped by operational maneuver. If there had been room to maneuver, the armies would have remained intact.

The largest battle of the war was fought at Gravelotte–Saint Privat, where 188,332 Prussians with 732 guns fought 112,800 French with 520 guns along a twelve-mile front. The size of the armies and the extent of the battlefront compared favorably with Wagram. The fire of the French forces at Gravelotte–Saint Privat was so great that although the French retreated from the field, they lost only 12,273 troops while inflicting 20,163 casualties on the Prussians.[14]

With the Franco-Prussian War, an era that began in 1809 closed. Although the advent of technological change altered warfare in the mid-nineteenth century in respect to strategy and tactics, the operational conduct of war remained remarkably consistent. The size of the armies remained about the same, as did their organization and the length of the battlefronts. Successful frontal assaults became quite rare, and armies could not be decisively defeated unless totally surrounded. Throughout this period, rival field armies retained a relational symmetry.

The period 1809–1871 was not static. There were changes wrought by technology, but the changes did not blot out many of the similarities. The next time the armies of Europe went to war, in the twentieth century, the nature of warfare had changed dramatically. The size of the armies, length of tactical fronts, and the advent of automatic weapons had altered war to such an extent as to make it unrecognizable from the warfare of the nineteenth century.

The critical point of this book is to convey the need for any student of warfare to understand the components that make up the practice of war at a given time. Those components must include the study of the opposing armies and how they interact in the dynamic of operations and combat. I have attempted to provide a holistic view of the Franco-Austrian War of 1809 and, by so doing, demonstrate that that war represents something greater than the sum of its separate parts. Consider the case of Napoleon. He built upon previous institutions and ideas to create a new type of warfare in 1805. The changes were decisive, considering the nature of his opposition. But the same man failed to realize that an entirely different approach to warfare was needed once his enemies modernized their armies.

Commentators and historians have blamed Napoleon's defeats after 1809 on personalities, in particular on mistakes made in judgment by Napoleon or his lieutenants during the conduct of operations. I have already discussed the so-called decline in Napoleon's abilities with respect to mistakes made in 1809, refuting this interpretation and arguing that the context of Napoleon's action had changed and that his abilities were really consistent. It has also been argued that Napoleon would have gotten his Austerlitz-style victory in Russia if his subordinate commanders, especially Jerome, had been more aggressive, or that a decisive victory would have been won at Bautzen in 1813 if only Marshal Ney had delayed in enveloping the Allied left. Macdonald, Ney, and Oudinot seemed incapable of acting effectively in independent command in 1813. Clausewitz spoke of friction and change in war as constant. One must accept the facts that if human nature is a constant, mistakes will be made on both sides in a war and that errors in judgment will be minimized or maximized depending on the relational dynamics of a given situation. Napoleon's mistakes prior to 1809 did not have catastrophic consequences because his opponents made more mistakes and their armies were less effective in war. But Napoleon's usual mistakes were magnified as his opposition improved. Convinced that his personality and genius could overcome all obstacles, Napoleon was blinded to the changing realities of warfare. Even today too much emphasis is placed on the roles that key individuals play in making heroic decisions at certain points in time. This creates the impression that to be effective in war requires only strong personality and the "will" to victory and implies that decisive victories can be won (the siren call of Austerlitz), as long as no mistakes are made. It was this approach to the use of history that led so many commanders in the nineteenth and the twentieth centuries to chase the image of Austerlitz or Cannae.

This book attempts to show that to understand the full relational dynamics of Napoleonic warfare, or warfare of any period, requires an understanding of the reasons for victory and defeat. One must examine the relational nature of command, the comparison of weapons, the structural resiliency, and the operational and tactical characteristics of any situation to get a realistic understanding of war. Studying only personalities skews the full picture of war, and the student will miss other elements worth noting. For the military officer, who studies military history to learn about his profession, the narrow drum-and-trumpet approach is particularly dangerous, as it leaves the impression that the outcome of war is determined by a few heroic men overcoming great odds. Perhaps this was Napoleon's view.

The serious student must understand all of the factors that make up the conduct of war in order to understand those factors in our own time and in the future. Although personalities and heroic decisions arc chronicled in this work, I have endeavored to show how all of the relational elements played a role in determining the course of the war of 1809 and the extent to which those elements continued into the nineteenth century. It is an approach that should be used for further study in the history of warfare.

Terms Used in the Text

Strategy refers to the overall plan of war. It includes the creation of theaters of operations, the assignment of objectives for the different theaters of operations, and the allocation of resources to the theaters of operations. *Operations, operational art,* and *the operational level of war* all refer to the plan and execution of campaigns designed to achieve strategic objectives within a theater of operations.

The term *manoeuvre sur les derrières* refers to envelopment. By this method an army would turn the flanks of an opposing force, sweep into its rear areas, and cut its lines of communication. Such a maneuver would disrupt the operations of the opposing force, compelling it either to disperse to regain its lines of communication or to turn the fight at a disadvantage to reopen them. It also means envelopment in a classic sense, when forces converge from different axes and surround an opponent. *Manoeuvre sur les derrières* is best used when an army is either equal or superior in numbers to its opponent.

The term *strategy of the central position* applies to the use of interior lines. This entails thrusting a concentrated army among the separate units of the enemy army and, by maneuvering within the interior lines gained by holding the central position, concentrating superior numbers successively against the detached units of the enemy.

Grand tactics refers to the overall plan of battle and deals with the movements of army corps and divisions on the battlefield. Various tasks would be assigned to army corps and divisions. They would serve either as the *masse primaire,* the *masse de manoeuvre,* or the *masse de rupture.* The role of the *masse primaire* was to engage frontally to pin down the enemy, preventing freedom of movement, and if possible to force the enemy to commit reserves prematurely. The *masse de manoeuvre* was a force designed to turn tactically either flank of the enemy, threatening its rear and rolling up its battleline. The *masse de rupture* (sometimes called the *masse de décision*) was a battle group held in reserve until, ideally, the enemy had used all of its reserves to fight the *masse primaire* and *masse de manoeuvre.* Once the enemy reserves were engaged, the *masse de*

Many of these terms are found in David Chandler, *The Campaigns of Napoleon* (New York: Macmillan, 1966), 165, 172, 1148–50.

rupture would be hurled at one point in the enemy line. The addition of this battle group to the fighting would produce a qualitative and quantitative superiority at that one point, forcing a decisive rupture of the enemy lines. The *masse de rupture* could also be used defensively as a reserve to blunt any enemy attack.

Tactics refers to the use of battalions, squadrons, and batteries within the corps or divisional structure to achieve their different grand tactical assignments. These tactical units were successively combined into regiments, brigades, divisions, and corps for easier administration and movement.

Le battalion carré literally means a battalion square. However, the term in the text refers to the operational deployment of an army into four major groupings consisting of a left wing, right wing, advance guard (sometimes called the center), and a reserve. Each of the component parts consists of one or more army corps. Each of the component parts are in supporting distance from the next, which is usually one to two days' march away. This deployment allows a large army to deploy over fronts thirty to seventy miles in length and move in almost any direction.

French Order of Battle

The following order of battle is a composite, as the names and composition of some of the units and commanders changed during the course of the war.

The Army of Germany

Commander in Chief, Emperor Napoleon I

Unit	Corps Commander	Division Commander or Commander of Attached Brigade
Imperial Guard		Curial, Dorsenne, Walther
II Corps	Oudinot, Lannes	Tharreau, Claparede, Saint-Hilaire or Grandjean, and Colbert's light cavalry brigade
III Corps	Davout	Morand, Friant, Gudin, Puthod, and Pajol's light cavalry brigade
IV Corps	Massena	Legrand, Carra Saint-Cyr, Molitor, Boudet, Lasalle, and Marulaz's light cavalry brigade
Cavalry Reserve Corps	Bessières	Nansouty, Saint-Sulpice, Espagne or Arrighi, Montbrun
VII Corps (Bavarian)	Lefebvre	Prince Ludwig, Deroi, Wrede, and light cavalry brigades under Zandt, Seydewitz, and Preysing
VIII Corps (Württemberger)	Vandamme	Neubronn, Wollwarth
IX Corps (Saxon)	Bernadotte, Reynier	Zezschwitz, Polenz, Dupas
X Corps (Westphalian)	King Jerome	Berchterode
Reserve Corps	Kellermann	Rivaud, Despeaux, Lagrange, Beaumont

Army of Italy

Commander in Chief, Prince Eugene de Beauharnais

Unit	Corps Commander	Division Commander
Royal Guard		Lecchi
Light Division		Dessaix
V Corps[a]	Macdonald	Lamarque, Broussier
VI Corps[a]	Grenier	Pacthod, Serras, Durutte
XI Corps[b]	Marmont	Montrichard, Clauzel
XII Corps[a] (Italian)	Baraguey d'Hilliers	Rusca, Severoli, Lauriston's Badenese brigade
Cavalry Corps	Grouchy	Sahuc, Guerin, Pully

Army of the Grand Duchy of Warsaw

Commander in Chief, Prince Poniatowski

Unit	Division Commander
First Division	Zayonchek
Second Division	Dombrowski

[a]The V, VI, and XII Corps were also sometimes referred to as Right, Center, and Left or First, Second, and Third Corps of the Army of Italy
[b]Also referred to as the Army of Dalmatia

Sources: David Chandler, *The Campaigns of Napoleon* (New York: Macmillan, 1966), 1107; Scott Bowden and Charlie Tarbox, *Armies on the Danube 1809* (Chicago: Emperor's Press, 1989), 183–202; Robert M. Epstein, *Prince Eugene at War: 1809* (Arlington, Tex.: Empire Games Press, 1984), 33, 125; Manfried Rauchensteiner, *Die Schlacht bei Deutsch Wagram am 5. und 6. Juli: 1809* (Vienna: Militärwissenschaftliches Institut, 1977), 59–61; J. Tranie and J. C. Carmigniani, *Napoleon et l'Autriche* (Paris: Copernic, 1979), 234–35; and *France Militaire*, ed. A. Hugo (Paris: Chez Delloye, 1837), 4:171.

Austrian Order of Battle

This order of battle is a composite owing to the many changes of commanders and units during the war.

Imperial and Royal Main Army

Commander in Chief, Archduke Charles

Unit	Corps Commander	Division Commander
I Corps	Bellegarde	Vogelsang, Ulm, Fresnel
II Corps[a]	Kollowrat, Hohenzollern	Brady, Treunenfels, Klenau
III Corps[a]	Hohenzollern, Kollowrat	Lusignan, Vukassovich, Saint Julien
IV Corps	Rosenberg	Hohenlohe, Dedovich, Somariva
V Corps	Archduke Louis, Reuss	Lindenau, Reuss, Schustekh
VI Corps	Hiller, Klenau	Jellachich, Vincent, Kottulinsky
Reserve Corps[b]	Lichtenstein	d'Aspre, Prochaszka, Hessen-Homburg, Schwarzenberg

Army of Inner-Austria

Commander in Chief, Archduke John

Unit	Corps Commander	Division Commander
VIII Corps	Chasteler, Albert Giulay	A. Giulay, Frimont
IX Corps	Ignatius Giulay	Gorup, Wolfskehl, Knesevich

Operating in Poland

Unit	Corps Commander	Division Commander
VII Corps	Archduke Ferdinand	Mohr, Monda, Schauroth

[a]The II and III Corps switched numbers during the war.
[b]Originally there were units designated as the I Reserve Corps and II Reserve Corps under Lichtenstein and Kienmayer, respectively. The II Reserve Corps had only three brigades and was merged with Lichtenstein's corps after Eckmühl.

Source: Scott Bowden and Charlie Tarbox, *Armies on the Danube 1809* (Chicago: Emperor's Press, 1989), 76–82, 136–38, 203–8.

Note on Ranks of General Officers

In the text, the word *general* is used to denote one of several different types of high-ranking officers. The specific terms and ranks for *general* are listed below.

AUSTRIAN

Generalissimus (Generalissimo, which is a better known term, is used in the text)	commander in chief of all armies
Feldmarschall	senior army commander
Feldzeugmeister or General der Kavallerie	commanded field armies or corps
Feldmarschall-Leutnant	commanded corps or divisions
Generalmajor (an older term was General-Feldwachtmeister)	commanded brigades

FRENCH

Marshal	commanded armies and corps
Colonel General	a purely ceremonial title, not a military rank
Lieutenant General	not a rank in the French army but, rather, a term or title sometimes used to designate a special position or for Generals of Division who commanded corps
General of Division	commanded corps and divisions
General of Brigade	commanded brigades

NOTES

CHAPTER ONE. INTRODUCTION

1. The problem of using history to affirm doctrine is effectively described in Gary P. Cox, "Of Aphorisms, Lessons, and Paradigms: Comparing the British and German Official Histories of the Russo-Japanese War," *Journal of Military History* 56:13 (July 1992): 389–401.

2. Michael Howard, "The Use and Abuse of Military History," *Parameters* 11 (March 1981): 10–15.

3. This theme is stressed in Antoine H. Jomini, *The Life of Napoleon*, trans. H. W. Halleck, 4 vols. (New York: D. Van Nostrand, 1864); Carl von Clausewitz, *On War*, ed. and trans. Michael Howard and Peter Paret (Princeton: Princeton University Press, 1976), B. H. Liddell Hart, *The Ghost of Napoleon* (New Haven: Yale University Press, 1933). Further discussion of the exaltation of Napoleon's victories can be found in Lorenzo M. Crowell, "The Illusion of the Decisive Napoleonic Victory," *Defense Analysis* 4 (1988): 329–45.

4. The exception to this is of course made by British authors in their approach to Wellington and the British army on the Peninsula and at Waterloo.

5. Russell F. Weigley, *The American Way of War* (Bloomington: Indiana University Press, 1973), 312.

6. Michael Howard, "Forgotten Dimensions of Strategy," in *The Causes of War* (Cambridge: Harvard University Press, 1983), 101–16.

7. Edward Hagerman, *The American Civil War and the Origins of Modern Warfare* (Bloomington: Indiana University Press, 1988), xi.

8. Matthew Cooper, *The German Army* (New York: Stein and Day, 1987), 132.

9. David Chandler, *The Campaigns of Napoleon* (New York: Macmillan, 1966), 733.

10. This is the case in the major published histories such as Chandler, *Campaigns of Napoleon*; F. Lorraine Petre, *Napoleon and the Archduke Charles* (London: Arms and Armor Press, 1976); Vincent J. Esposito and John Robert Elting, *A Military History and Atlas of the Napoleonic Wars* (New York: Praeger, 1964); and Owen Connelly, *Blundering to Glory* (Wilmington, Del.: Scholarly Resources, 1984).

11. For example, David Chandler writes that Napoleon's eventual defeat "was partly due to Napoleon's failure to train up his subordinates for the exigencies of independent command" (*Campaigns of Napoleon*, 932).

CHAPTER TWO. THE TRANSFORMATION OF WARFARE

1. Martin Van Creveld, in *Command in War* (Cambridge: Harvard University Press, 1985), 58, refers to this as "the Revolution in Strategy."

2. There were, of course, exceptions, such as the 110,000-man Allied army at Malplaquet.

3. Maurice de Saxe, *Reveries on the Art of War*, trans. Thomas R. Phillips (Harrisburg, Pa.: Military Service Publishing Company, 1953), 36–38, 50–51.

4. R. S. Quimby, *The Background of Napoleonic Warfare* (New York: Columbia University Press, 1957), 94.

5. Pierre de Bourcet, *Principes de la guerre de montagnes* (Paris: Imprimerie National, 1877). The concept of branches and flexibility is stressed throughout the book. In particular, this concept is discussed on pp. 88–90 and 92–93.

6. The term *operational level* is a relatively new one and is used to describe the conduct of war between the strategic and tactical levels. It includes the attempt to achieve strategic goals by the application of military forces and the development of campaign plans in a theater of operations. As strategy determines operations, the operational level drives tactical situations.

7. Quimby, *Background of Napoleon's Warfare*, 113.

8. Ibid.

9. Steven T. Ross, *European Diplomatic History, 1789–1815* (Garden City, N.Y.: Doubleday, 1969), 80. (Record keeping was inadequate, and these are paper strengths. In reality, they were somewhat fewer and frontline strength was probably at 300,000.)

10. Ibid., 84.

11. There is an ongoing debate about the extent of the use of columns and skirmish tactics in the revolutionary armies. Among the best and most recent studies on the subject is John A. Lynn, *The Bayonets of the Republic* (Urbana: University of Illinois Press, 1984).

12. Napoleon I, *La Correspondance de Napoléon Ier* (Paris: Henri Plon and J. Dumaine, 1861), 16:154, hereafter cited as NC.

13. Ibid., 201.

14. David Chandler, *The Campaigns of Napoleon* (New York: Macmillan, 1966), 266.

15. Vincent J. Esposito and John Robert Elting, *A Military History and Atlas of the Napoleonic Wars* (New York: Praeger, 1964), 35.

16. Ibid.

17. Ibid.

18. Chandler, *Campaigns of Napoleon*, 267.

19. Ibid., 268–69.

20. Ibid., 269.

21. George Armand Furse, *1800 Marengo and Hohenlinden* (London: William Coles, 1903), 458.

22. Russell F. Weigley, *The Age of Battles* (Bloomington: Indiana University Press, 1991) 373.

23. The best account in English of Charles and the functions of the Hofkriegsrat can be found in Gunther E. Rothenberg, *Napoleon's Great Adversaries: The Archduke Charles and the Austrian Army 1792–1814* (Bloomington: Indiana University Press, 1982).

24. Ibid., 72.

25. Ibid., 78.

26. Ibid., 85–88.

27. Ibid., 85.

28. Ibid., 74.

29. Chandler, *Campaigns of Napoleon*, 1102.

30. Ibid., 1103.

31. Paul Claude Alombert and Jean Lambert Alphonse Colin, *La Campagne de 1805 en Allemagne*, 6 vols. (Paris: Librairie Militaire R. Chapelot, 1902–1908) 4:726–30.

32. Esposito and Elting, *Military History and Atlas*, 47.

33. *NC*, 11:282.

34. Christopher Duffy, *Austerlitz* (London: Seely, 1977), 181–84.

35. Ibid., 161.

36. Ibid., 156–57.

37. J. D. Hittle, *The Military Staff* (Westport, Conn.: Greenwood, 1975), 58–63.

38. Esposito and Elting, *Military History and Atlas*, 68.

39. Gunther E. Rothenberg, *The Art of Warfare in the Age of Napoleon* (Bloomington: Indiana University Press, 1978), 201.

CHAPTER THREE. ARMIES FOR GERMANY

1. Gunther E. Rothenberg, *Napoleon's Great Adversaries: The Archduke Charles and the Austrian Army 1792–1814* (Bloomington: Indiana University Press, 1982). This is a major theme in this book.

2. Ibid., 106.

3. Ibid., 108.

4. Ibid., 108–9.

5. Ibid., 118.

6. Ibid., 119.

7. David Chandler, *The Campaigns of Napoleon* (New York: Macmillan, 1966), 668.

8. Rothenberg, *Napoleon's Great Adversaries*, 122.

9. Chandler, *Campaigns of Napoleon*, 668–69.

10. Robert M. Epstein, *Prince Eugene at War: 1809* (Arlington, Tex.: Empire Games Press, 1984), 19.

11. Rothenberg, *Napoleon's Great Adversaries*, 121.

12. Ibid., 115.

13. Ibid., 127.

14. The latest and best work on the modernization and combat performance of the armies of the Confederation of the Rhine is John H. Gill, *With Eagles to Glory: Napoleon and His German Allies in the 1809 Campaign* (Novato, Calif.: Presidio Press, 1992).

15. Ibid., 256.

16. Chandler, *Campaigns of Napoleon*, 1107.

17. Ibid., and Scott Bowden and Charlie Tarbox, *Armies on the Danube 1809* (Chicago: Emperor's Press, 1989), 68–75.

18. Bowden and Tarbox, *Armies on the Danube 1809*, 72–73.

19. Ibid., 72–74.

20. *NC*, 18:403–6.

21. Ibid., 407–8.

22. F. Lorraine Petre, *Napoleon and the Archduke Charles* (London: Arms and Armor Press, 1976), 61.

23. Ibid., 691.

CHAPTER FOUR. ARMIES FOR ITALY

1. *NC*, 18:247.

2. Ibid.

3. Eugene de Beauharnais, *Memoires et correspondence politique et militaire du Prince Eugène*, ed. Andre du Casse, 10 vols. (Paris: Michel Lévy Freres, 1859), 4:243–44 (hereafter cited as *DCC*. There is also a biographical sketch included in these volumes.

4. *NC*, 18:213–14.

5. Ibid.

6. Ibid., 217.

7. Ibid., 217–18.

8. Ibid., 218.

9. *DCC*, 4:308–13.

10. See chapters 3 and 5.

11. Scott Bowden and Charlie Tarbox, *Armies on the Danube 1809* (Chicago: Emperor's Press, 1989), 138.

12. *DCC*, 4:446.

13. Ibid., 369–70, 377.

14. Bowden and Tarbox, *Armies on the Danube 1809*, 132.

15. Ibid., 134.

16. *DCC*, 4:366.

17. Ibid., 413.

18. Ibid., 413–14.

19. *NC*, 18:402. Napoleon used the term *lieutenant general* for the specified commanders, which was a designation of function and not a formal rank in the French army at this time. The men nominated held the rank of general of division.

20. Ibid., 404.

21. *DCC*, 4:358–59, 401–2.

22. Ibid., 399–400, 444–49.

CHAPTER FIVE. WAR ALONG THE DANUBE

1. David Chandler, *The Campaigns of Napoleon* (New York: Macmillan, 1966), 670.

2. Vincent J. Esposito and John Robert Elting, *A Military History and Atlas of the Napoleonic Wars* (New York: Praeger, 1964), 93.

3. Notice the increase in the proportion of artillery. In 1805, the *Grande Armée* had 210,000 troops and 395 guns.

4. F. Lorraine Petre, *Napoleon and the Archduke Charles* (London: Arms and Armor Press, 1976), 76.

5. Chandler, *Campaigns of Napoleon*, 677–78. This account of the missing message is also told in Petre, *Napoleon and Archduke Charles*, 77–85, and James R. Arnold, *Crisis on the Danube* (New York: Paragon, 1990), 59–61.

6. Esposito and Elting, *Military History and Atlas*, 97.

7. Chandler, *Campaigns of Napoleon*, 689.

8. Ibid., 693.

9. *DCC*, 5:150–52, 156–57.

CHAPTER SIX. CRISIS IN ITALY

1. *DCC*, 4:329.

2. Ibid., 397.

3. Ibid., 435.

4. Ibid., 371–72.

5. Ibid., 392.

6. Manuscript, Beauharnais Collection, Princeton University, Box 27, Letter Book 1809 (hereafter cited as MSP, 1809).

7. *DCC*, 4:415.

8. Ibid., 342.

9. Ibid., 281.

10. MSP, Box 27, L.B. 1809.

11. *NC*, 18:404.

12. *DCC*, 4:442.

13. Ibid., 440–41.

14. Frederick-François-Guillaume, Baron de Vaudoncourt, *Histoire politique et militaire du Prince Eugène Napoléon*, 2 vols. (Paris: Librairie Universelle de P. Mongie, 1828), 1:138–39 (hereafter cited as Vaud).

15. Attributed to Moltke.

16. *DCC*, 4:446–48.

17. Ibid., 135–36.

18. *DCC*, 5:134–35.

19. Ibid., 135–36.

20. For a more detailed account of this battle and the other operations of the Army of Italy, see Robert M. Epstein, *Prince Eugene at War: 1809* (Arlington, Tex.: Empire Games Press, 1984).

21. *DCC*, 5:143–74.

22. Ibid., 137.

23. Vaud., 1:197.

24. MSP, Box 27, L.B. 1809.

25. Ibid.

26. Pacthod would later command this division.

27. This had been Severoli's division. Wounded at Sacile, he would return to command later.

28. The sources vary for the exact nomenclature of these corps. The correspondence tends to refer to them under the name of the respective corps commander. Jean-Jacques-Germain Pelet, *Memoires sur la guerre de 1809*, 4 vols. (Paris: Librairie Roret, 1825) uses the term left, center, and right for each corps. L. Loy, *La campagne de Styria en 1809* (Paris: Librairie Militaire, R. Chapelot, 1908) and J. Tranie and J. C. Carmigniani, *Napoléon et l'Autriche* (Paris: Copernic, 1979), 235, indicate that the corps were numbered 1, 2, and 3 of the Army of Italy. Manfried Rauchensteiner, *Die Schlacht bei Duetsch Wagram am 5. und 6. Juli 1809* (Vienna: Militärwissenschaftliches Institute, 1977), 59, indicates that Macdonald's and Grenier's corps were numbered V and VI, respectively. I am inclined to follow his terminology, because the corps in Napoleon's Army of Germany were numbered II through IV for the French corps and began at VII for the German corps. Macdonald's and Grenier's corps were both French, so designating them V and VI makes up the gap between Napoleon's IV and VII Corps in the order of battle. The German corps were numbered VII through X, and Marmont's Army of Dalmatia was eventually renumbered XI. I am assuming that Baraguey d'Hilliers' corps became the XII Corps, being the last corps in the order of battle. It is possible that the corps of the Army of Italy were renumbered later in the war when they joined Napoleon's forces in Austria. To avoid confusion, the corps of the Army of Italy will be designated in this study either by the names of their commanders or by the Roman numerals V, VI, XI, and XII. For the cavalry attachments to the V and VI Corps, see *France militaire: Histoire des Armées Français de terre et de mer de 1792 1833*, ed. A. Hugo, (Paris: Chez Delloye, 1837), 4:171

29. This is not to be confused with the left, center, and right referred to as tactical terms for armies of the previous century.
30. *DCC*, 5:150–52.
31. Ibid., 156–57.
32. Ibid., 157–60.

CHAPTER SEVEN. VICTORY IN ITALY

1. Vaud., 1:186–87.
2. Ibid., 210.
3. *DCC*, 5:165.
4. David Chandler, *The Campaigns of Napoleon* (New York: Macmillan, 1966), 275.
5. MSP, Box 27, L.B. 1809.
6. Robert M. Epstein, *Prince Eugene at War: 1809* (Arlington, Tex.: Empire Games Press, 1984), 830.
7. Ibid., 90.
8. Ibid., 93.
9. Ibid., 94.
10. Epstein, *Prince Eugene*, 94.
11. *DCC*, 5:161.
12. French estimates are in *DCC*, 5:190–92; Austrian figures are in Gaston Bodart, *Militär-historisches Kriegs-Lexikon* (Vienna and Leipzig: Stern, 1908), 402.
13. Vaud. 1:264.

CHAPTER EIGHT. MARCH ON VIENNA AND THE BATTLE OF
ASPERN-ESSLING

1. *DCC*, 5:150–52, 156–57.
2. The II Corps and Lannes' provisional force were reorganized. The provisional task force was broken up, and some of the divisions went back to Davout while Lannes was given Saint-Hilaire's infantry division. As part of Lannes' II Corps, Oudinot would command the Corps of Grenadiers, consisting of the infantry divisions of Tharreau and Claparede.
3. F. Lorraine Petre, *Napoleon and the Archduke Charles* (London: Arms and Armor Press, 1976), 241.
4. Gunther E. Rothenberg, *Napoleon's Great Adversaries: The Archduke Charles and the Austrian Army 1792–1814* (Bloomington: Indiana University Press, 1982), 136.
5. In 1866, Benedek came to the same realization about his army. He abandoned mobile operations against the Prussians and concentrated his army for a defensive positional battle at Sadowa.
6. The only actual change was that the corps were renamed *columns*. The term *corps* will continue to be used in the text.
7. John H. Gill, *With Eagles to Glory: Napoleon and His German Allies in the 1809 Campaign* (Novato, Calif.: Presidio Press, 1992), 335.
8. David Chandler, *The Campaigns of Napoleon* (New York: Macmillan, 1966), 696–97.
9. Scott Bowden and Charlie Tarbox, *Armies on the Danube 1809* (Arlington, Tex.: Empire Games Press, 1980), 84.

10. Chandler, *Campaigns of Napoleon*, 703.
11. Petre, *Napoleon and the Archduke Charles*, 264.
12. Bowden and Tarbox, *Armies on the Danube*, 93.
13. Vincent J. Esposito and John Robert Elting, *A Military History and Atlas of the Napoleonic Wars* (New York: Praeger, 1964), 101.
14. Petre, *Napoleon and the Archduke Charles*, 272.
15. Ibid., 282.
16. Bowden and Tarbox, *Armies on the Danube*, 83.
17. Ibid., 86.
18. Petre, *Napoleon and the Archduke Charles*, 284.
19. Chandler, *Campaigns of Napoleon*, 702.
20. Ibid., 703.
21. Petre, *Napoleon and the Archduke Charles*, 294.
22. Ibid., 296.
23. Ibid., 298.

CHAPTER NINE. JUNCTION

1. Vaud., 1:259–60.
2. Grenier had the infantry divisions of Durutte and his old division, now commanded by General of Division Pacthod. Baraguey d'Hilliers' corps had only Fontanelli's division.
3. Vaud., 1:276.
4. Robert M. Epstein, *Prince Eugene at War: 1809* (Arlington, Tex.: Empire Games Press, 1984), 102–3.
5. Vaud., 1:242.
6. *DCC*, 5:212–24.
7. Ibid., 214–15.
8. For accounts of the casualties see MSP, Box 27, L.B. 1809; *DCC*, 5:228–32; Vaud., 1:303–9; L. Loy, *La Campagne de Styrie en 1809* (Paris: Librairie Militaire, R. Chapelot, 1908), 9–17; and Epstein, *Prince Eugene*, 107–8.
9. Vaud., 1:293–96.

CHAPTER TEN. THE WAGRAM CAMPAIGN: FIRST PHASE

1. *NC*, 19:50.
2. Ibid., 44.
3. Ibid., 57.
4. Ibid., 51.
5. F. Lorraine Petre, *Napoleon and the Archduke Charles* (London: Arms and Armor Press, 1976), 352.
6. *DCC*, 5:379–80.
7. Vaud., 1:349.
8. Estimates for Giulay's corps vary during June. In the early part of the month, Eugene estimated Giulay to have from 7,000 to 8,000 troops (*DCC*, 5:353–54). Vincent J. Esposito and John Robert Elting, *A Military History and Atlas of the Napoleonic Wars* (New York: Praeger, 1964), 103, estimate Giulay had 19,000. Giulay was reinforced during June, and by the end of the month had 15,000, of which a large proportion were of little military value.
9. *DCC*, 5:315.

10. Ibid.
11. Ibid., 334.
12. Ibid., 334–37.
13. Ibid., 315.
14. Ibid., 338.
15. This corps consisted of Montbrun's division and Grouchy's former dragoon division now commanded by General of Brigade François Guerin d'Etoquigny.
16. *DCC*, 5:362.
17. Ibid., 360–62.
18. Ibid., 349.
19. Vaud., 1:349.
20. Scott Bowden and Charlie Tarbox, *Armies on the Danube 1809* (Arlington, Tex.: Empire Games Press, 1980), 123.
21. Vaud., 1:346.
22. Petre, *Napoleon and the Archduke Charles*, 310.
23. Vaud., 1:349–50.
24. Ibid., 345.
25. Bowden and Tarbox, *Armies on the Danube 1809*, 120.
26. *DCC*, 5:377–78.
27. The following account of the Battle of Raab is based on MSP, 27. L.B. 1809; *DCC*, 5:377–84; Vaud., 1:352–68; and Jean-Jacques-Germain Pelet, *Mémoires sur la guerre de 1809*, 4 vols. (Paris: Librairie Roret, 1825), 4:94–104.
28. Two horse artillery batteries had been attached to each cavalry division.
29. *DCC*, 5:382–83.
30. Ibid., 379–80.
31. Ibid., 406.
32. Gaston Bodart, *Militär-historisches Kriegs-Lexikon* (Vienna and Leipzig: Stern, 1908), 408.
33. Ibid., 449–50

CHAPTER ELEVEN. THE WAGRAM CAMPAIGN: SECOND PHASE

1. David Chandler, *The Campaigns of Napoleon* (New York: Macmillan, 1966), 708.
2. Ibid., 709.
3. Ibid.
4. F. Lorraine Petre, *Napoleon and the Archduke Charles* (London: Arms and Armor Press, 1976), 352.
5. Gunther E. Rothenberg, *Napoleon's Great Adversaries: The Archduke Charles and the Austrian Army 1792–1814* (Bloomington: Indiana University Press, 1982), 161.
6. Baron de Marbot, *The Memoirs of Baron de Marbot*, trans. Arthur John Butler (London and New York: Longmans, Green and Company, 1893), 367.
7. *DCC*, 5:465–67.
8. *NC*, 19:205–8.
9. Petre, *Napoleon and the Archduke Charles*, 333.
10. Ibid.
11. Scott Bowden and Charlie Tarbox, *Armies on the Danube 1809* (Arlington, Tex.: Empire Games Press, 1980), 154.

12. Vincent J. Esposito and John Robert Elting, *A Military History and Atlas of the Napoleonic Wars* (New York: Praeger, 1964), 104.

13. Argued by Petre and others.

14. This was formerly Saint-Hilaire's command.

15. Order of battle from Bowden and Tarbox, *Armies on the Danube 1809*, 140–54.

16. Vaud, 1:402–3.

17. Ibid., 405–6.

18. Petre, *Napoleon and the Archduke Charles*, 361.

19. Bowden and Tarbox, *Armies on the Danube 1809*, 148–49.

20. Etienne-Jacques-Joseph Alexandre Macdonald, *The Recollections of Marshal Macdonald*, ed. Camille Rousset, trans. Stephen Louis Simeon (New York: Charles Scribner's Sons, 1893), 334.

21. Rothenberg, *Napoleon's Great Adversaries*, 169.

22. Chandler, *Campaigns of Napoleon*, 722–23.

23. Marbot, *Memoirs*, 386.

24. Chandler, *Campaigns of Napoleon*, 723.

25. Marbot, *Memoirs*, 386.

26. Bowden and Tarbox, *Armies on the Danube 1809*, 168.

27. Ibid., 154.

28. Vaud., 1:409.

29. Chandler, *Campaigns of Napoleon*, 728.

30. Marbot, *Memoirs*, 383.

31. Vaud., 1:413.

32. Jean-Jacques-Germain Pelet, *Memoires sur la guerre de 1809*, 4 vols. (Paris: Librairie Roret, 1825), 4:222.

33. Chandler, *Campaigns of Napoleon*, 728. A formation similar to this was proposed by Napoleon in consultation with Marshal Soult when preparing the assault on the Pratzen Heights at Austerlitz in 1805—when the imperial infantry was reputedly at its height.

34. Vaud., 1:415.

35. Petre, *Napoleon and the Archduke Charles*, 380. It would not be unusual for the victor not to know the actual outcome of so vast a battle until the following day. The same phenomenon occurred at Gravelotte-Saint Privat in 1870.

36. Petre, *Napoleon and the Archduke Charles*, 363.

37. Rothenberg, *Napoleon's Great Adversaries*, 168.

38. Vaud., 1:416.

39. Ibid.

40. Rothenberg, *Napoleon's Great Adversaries*, 168.

41. Petre, *Napoleon and the Archduke Charles*, 378.

42. Ibid., 379.

43. This issue had been studied in Robert M. Epstein, *Prince Eugene At War: 1809* (Arlington, Tex.: Empire Games Press, 1984). Macdonald, *Recollections*, plays up the idea that he was really commanding the Army of Italy. However, the correspondence of both Napoleon and Eugene give no such interpretation; Eugene de Beauharnais commanded the Army of Italy, not Macdonald.

44. Owen Connelly, ed., *Historical Dictionary of Napoleonic France, 1799–1815* (Westport, Conn.: Greenwood Press, 1985), 496.

45. *NC*, 19:361.

46. Petre, *Napoleon and the Archduke Charles*, 361.

47. Chandler, *Campaigns of Napoleon*, 731–32.

48. John G. Gallaher, *The Iron Marshal* (Carbondale: Southern Illinois University Press, 1976), 202.

CHAPTER TWELVE. THE EMERGENCE OF MODERN WAR

1. Those works able to describe Napoleonic practice from his actions and correspondence include the following works by Hubert Camon: *La Guerre napoléonienne*, 5 vols. (Paris: Chapelot, 1903–1910); *La systeme de guerre de Napoléon* (Paris: Berger-Levrault, 1923); *Quand et comment Napoléon a conçu son system de manoeuvre* (Paris: Berger-Levrault, 1931); and *Quand et comment Napoléon a conçu son systeme de bataille* (Paris: Berger-Levrault, 1935). In addition, there are Jean Lambert Colin, *Les transformations de la guerre* (Paris: Flammarion, 1911), and David Chandler, *The Campaigns of Napoleon* (Macmillan, New York: 1966).

2. Chandler, *Campaigns of Napoleon*, 733.

3. For an elaboration on this theme, see, Robert M. Epstein, "Patterns of Change and Continuity in Nineteenth Century Warfare," *Journal of Military History* 56:3 (July 1992): 375–88.

4. Chandler, *Campaigns of Napoleon*, 1114.

5. Vincent J. Esposito and John Robert Elting, *A Military History and Atlas of the Napoleonic Wars* (New York: Praeger, 1964), 116.

6. Epstein, "Patterns," 382.

7. Ibid., 384.

8. Chandler, *Campaigns of Napoleon*, 1120.

9. Ibid., 1120–21.

10. Epstein, "Patterns," 386.

11. Paddy Griffith, *Forward into Battle: Fighting Tactics from Waterloo to the Near Future* (Novato, Calif.: Presidio Press, 1981), 78.

12. Epstein, "Patterns," 386–87.

13. Ibid., 389.

14. Ibid., 388.

BIBLIOGRAPHY

PRIMARY SOURCES

Manuscripts

Beauharnais Collection. Princeton Archives. Princeton University.

Published Documents

Beauharnais, Eugene de. *Memoires et correspondance politique et militaire du Prince Eugène.* Edited by Andre du Casse. 10 vols. Paris: Michel Lévy Fréres, 1859. (Included in this collection is a biography written by Andre du Casse.)
France militaire. Histoire des armée français de terre et de mer de 1792–1833. Edited by A. Hugo. 5 vols. Paris: Chez Delloye, 1837.
Napoleon I. *La Correspondance de Napoléon Ier.* 36 vols. Paris: Henri Plon and J. Dumaine, 1858–1870.

Memoirs and Histories

Bonaparte, Hortense de Beauharnais. *The Memoirs of Queen Hortense.* Edited by Jean Hanoteau. 2 vols. New York: Cosmopolitan Book Corporation, 1928.
Darnay, Baron. *Notices historiques sur S.A.R. Le Prince Eugène Vice-roi d'Italie.* Paris: David, 1830.
Jomini, Antoine H. *The Life of Napoleon.* Translated by H. W. Halleck. 4 vols. New York: D. Van Nostrand, 1864.
Macdonald, Etienne-Jacques-Joseph Alexandre. *The Recollections of Marshal Macdonald.* Edited by Camille Rousset; translated by Stephen Louis Simeon. New York: Charles Scribner's Sons, 1893.
Marbot, Baron de. *The Memoirs of Baron de Marbot.* Translated by Arthur John Butler. London and New York: Longmans, Green and Company, 1893.
Marmont, August F. L. V. *Memoires du Marechal Marmont Duc de Raguse.* 9 vols. Paris: Perrotine, 1857.
Massena, Andre. *Memoires de Masséna.* Edited by General J. B. F. Koch. 7 vols. Paris: Paulin and Lechevalire, 1848–1850.
Pelet, Jean-Jacques-Germain. *Memoires sur la guerre de 1809,* 4 vols. Paris: Librairie Roret, 1825.
Vaudoncourt, Frederick-Francois-Guillaume, Baron de. *Histoire politique et militaire du Prince Eugène Napoléon.* 2 vols. Paris: Librairie Universelle de P. Mongie, 1828.

SECONDARY SOURCES

Adalbert, Prinz von Bayern. *Eugene Beauharnais, der Stiefsohn Napoleons.* Munich: F. Bruckmann, 1940.
Alombert, Paul Claude, and Jean Lambert Colin. *La Campagne de 1805 en Allemagne.* 6 vols. Paris: Librairie Militaire R. Chapelot, 1902–1908.
Arnold, James R. *Crisis on the Danube.* New York: Paragon, 1990.
Bernady, Françoise. *Eugène de Beauharnais.* Paris: Library Academique Perrin, 1973.
Bodart, Gaston. *Militär-historisches Kriegs-Lexikon.* Vienna and Leipzig: Stern, 1908.
Bonnal, Henri G. *La Manoeuvre de Landshut.* Paris: Chapolot, 1905.
Bourcet, Pierre de. *Principes de la guerre de montagnes.* Paris: Imprimerie National, 1877.
Bowden, Scott. *Napoleon's Grande Armée of 1813.* Chicago: Emperor's Press, 1990.
Bowden, Scott, and Charlie Tarbox. *Armies on the Danube 1809.* Arlington, Tex.: Empire Games Press, 1980. Reprinted Chicago: Emperor's Press, 1989.
Bruun, Geoffrey. *Europe and the French Imperium.* New York: Harper and Row, 1965.
Camon, Hubert. *La Guerre napoléonienne,* 5 vols. Paris: Chapelot, 1903–1910.
———. *Quand et comment Napoléon a conçu son systeme de bataille.* Paris: Berger-Levrault, 1935.
———. *Quand et comment Napoléon a conçu son systeme de manoeuvre.* Paris: Berger-Levrault, 1921.
———. *La Systeme de guerre de Napoléon.* Paris: Berger-Levrault, 1923.
Cassi, Gellio. "Napoleon et la defense de l'Italie sur la Piave." *Revue des etudes napoléoniennes.* 19 (July-August 1923): 5–23. (Includes a reprint of an eyewitness account of the Battle of Sacile.)
Chandler, David. *The Campaigns of Napoleon.* New York: Macmillan, 1966.
Chandler, David, ed. *Napoleon's Marshals.* New York: Macmillan, 1987.
Clausewitz, Carl von. *On War.* Edited and translated by Michael Howard and Peter Paret. Princeton: Princeton University Press, 1976.
Colin, Jean Lambert. *Les transformations de la guerre.* Paris: Flammarion, 1911.
Connelly, Owen. *Blundering to Glory.* Wilmington, Del.: Scholarly Resources, 1984.
Connelly, Owen, ed. *Historical Dictionary of Napoleonic France, 1799–1815.* Westport, Conn.: Greenwood Press, 1985.
Cooper, Matthew. *The German Army.* New York: Stein and Day, 1978.
Cox, Gary P. "Of Aphorisms, Lessons, and Paradigms: Comparing the British and German Official Histories of the Russo-Japanese War," *Journal of Military History* 56 (July 1992): 389–401.
Crowell, Lorenzo M. "The Illusion of the Decisive Napoleonic Victory." *Defense Analysis* 4 (1988): 329–45.
Delderfield, R. F. *Napoleon's Marshals.* Philadelphia and New York: Chilton, 1962.
Driault, Edouard. *Napoleon en Italie (1800–1812).* Paris: Felix Alcan, 1906.
Duffy, Christopher. *Austerlitz.* London: Seeley Service, 1977.
———. *Borodino.* New York: Charles Scribner's Sons, 1973.
Elting, John R. *Swords around a Throne.* New York: Free Press, 1988.
Epstein, Robert M. "Patterns of Change and Continuity in Nineteenth Century Warfare." *Journal of Military History* 56:3 (July 1992): 375–88.

———. *Prince Eugene at War: 1809*. Arlington, Tex.: Empire Games Press, 1984.

Esposito, Vincent J., and John Robert Elting. *A Military History and Atlas of the Napoleonic Wars*. New York: Praeger, 1964.

Foch, Ferdinand. *The Principles of War*. Translated by J. de Morinni. New York: H. K. Fly Company, 1918.

Fugier, Andre. *Napoleon et l'Italie*. Paris: J. B. Janin, 1947.

Furse, George Armand. *1800 Marengo and Hohenlinden*. London: William Coles, 1903.

Gachot, Edouard. *1809 Napoleon en Allemagne*. Paris: Plon-Nourrit, 1913.

———. *La Troisiéme campagne de'Italie 1805–1806*. Paris: Plon-Nourrit, 1911.

Gallaher, John G. *The Iron Marshal*. Carbondale: Southern Illinois University Press, 1976.

Gallois, N. *Armées francaises en Italie 1494–1849*. Paris: A. Bourdillait, 1859.

Gill, John H. *With Eagles to Glory: Napoleon and his German Allies in the 1809 Campaign*. Novato, Calif.: Presidio Press, 1992.

Glover, Michael. *The Peninsular War*. Hamden, Conn.: Archon Books, 1966.

Griffith, Paddy. *Forward into Battle: Fighting Tactics from Waterloo to the Near Future*. Novato, Calif.: Presidio Press, 1981.

Hagerman, Edward. *The American Civil War and the Origins of Modern Warfare*. Bloomington: Indiana University Press, 1988.

Haig, Douglas. *Cavalry Studies—Strategical and Tactical*. London: Hugh Rees, 1907.

Hittle, J. D. *The Military Staff*. Westport, Conn.: Greenwood, 1975. Originally published Harrisburg, Pa.: Military Service Publishing Company, 1944.

Holtman, Robert B. *The Napoleonic Revolution*. Philadelphia and New York: J. B. Lippincott, 1967.

Horward, Donald D. *Napoleonic Military History: A Bibliography*. New York: Garland, 1986.

Horward, Donald D., and John C. Horgan, eds. *The Consortium on Revolutionary Europe: Proceedings, 1989*. Tallahassee, Fla.: Rose, 1990.

Howard, Michael. *The Causes of War*. Cambridge: Harvard University Press, 1983.

———. "The Use and Abuse of Military History." *Parameters*. 11 (March 1981): 10–15.

Humble, Richard. *Napoleon's Peninsular Marshals*. New York: Taplinger, 1973.

Lefebvre, Georges. *The French Revolution*. Vol. 1 translated by Elizabeth Moss Evanson; vol. 2 translated by John Hall and James Friguglieti. New York: Columbia University Press, 1962.

———. *Napoleon*. Vol. 1, translated by Henry F. Stockhold; vol. 2 translated by J. E. Anderson. New York: Columbia University Press. 1969.

Levy, Arthur. *Napoleon et Eugène*. Paris: Calmann-Levy, 1926.

Liddell Hart, B. H. *The Ghost of Napoleon*. New Haven: Yale University Press, 1933.

Loy, L. *La Campagne de Styria en 1809*. Paris: Librairie Militaire, R. Chapelot, 1908.

Lynn, John A. *The Bayonets of the Republic*. Urbana: University of Illinois Press, 1984.

McDonough, James R. *The Limits of Glory*. Novato, Calif.: Presidio Press, 1991.

Markham, Felix. *Napoleon*. New York: Mentor, 1963.

———. *Napoleon and the Awakening of Europe*. New York: Collier, 1965.

Marshall-Cornwall, Sir J. *Marshal Massena*. London: Oxford University Press, 1965.

Masson, Frederic. *Napoleon et sa famille*. Paris: Paul Ollendorff, 1900.

Montagu, V. M. *Napoleon and His Adopted Son.* New York: McBride, Nase and Company, 1914.

Oman, Carola. *Napoleon's Viceroy.* London: Hodder and Stoughton, 1966.

Parker, Harold T. *Three Napoleonic Battles.* Durham, N.C.: Duke University Press, 1944.

Petre, F. Lorraine. *Napoleon and the Archduke Charles.* London: Arms and Armor Press, 1976. Originally published London: John Lane, 1909.

Phipps, Colonel R. W. *The Armies of the First French Republic.* 5 vols. London: Oxford University Press, 1935–1939.

Pingaud, A. "Le premier royaume d'Italie; le developpemend du systeme Napoléonien." *Revue des etudes napoléoniennes* 21 (July-December, 1923), 34–50 and 100–110.

Pulitzer, Albert. *The Romance of Prince Eugene.* 2 vols. Translated by B. M. Sherman. New York: Dodd and Company, 1895.

Quennevat, Jean Claude. *Atlas de la Grande Armée.* Paris: Editions Sequoia, 1966.

Quimby, R. S. *The Background of Napoleonic Warfare.* New York: Columbia University Press, 1957.

Rauchensteiner, Manfried. *Die Schlacht bei Deutsch Wagram am 5. und 6. Juli 1809.* Militärhistorische Schriftenreihe Herausgegeben vom Heeresgeschichtlichen Museum. Vienna: Militärwissenschaftliches Institut, 1977.

Ross, Steven T. *European Diplomatic History, 1789–1815.* Garden City, N.Y.: Doubleday, 1969.

Rothenberg, Gunther E. *The Army of Francis Joseph.* West Lafayette, Ind.: Purdue University Press, 1976.

————. *The Art of Warfare in the Age of Napoleon.* Bloomington and London: Indiana University Press, 1978.

————. *Napoleon's Great Adversaries: The Archduke Charles and the Austrian Army 1792–1814.* Bloomington: Indiana University Press, 1982.

Saski, C. G. L. *Campagne de 1809 en Allemagne et Autriche.* 3 vols. Paris: Berger-Levrault, 1899.

Saxe, Maurice de. *Reveries on the Art of War.* Translated by Thomas R. Phillips. Harrisburg, Pa.: Military Service Publishing Company, 1953.

Schneider, James J. *Mars Ascending: Total War and the Rise of the Soviet Warfare State 1864–1929.* Ann Arbor: University Microfilms, 1992.

Six, Georges. *Dictionnaire biographique des géneraux et amiraux de la Révolution et de l'Empire.* 2 vols. Paris: G. Saffroy, 1934.

————. *Les Generaux de la Révolution et d l'Empire.* Paris: Bordus, 1947.

Thiers, Adolphe M. *Histoire du consulat et de l'Empire,* 21 vols. Paris: Michel Lévy Fréres, 1845–1869.

Tranie, J., and J. C. Carmigniani. *Napoléon et l'Autriche.* Paris: Copernic, 1979.

Van Creveld, Martin. *Command in War.* Cambridge: Harvard University Press, 1985.

Weigley, Russell F. *The Age of Battles.* Bloomington: Indiana University Press, 1991.

————. *The American Way of War.* Bloomington: Indiana University Press, 1973.

INDEX